D1356456

WP 0153127 1
574.876/REC

Receptors and Recognition

Series A

Published

Volume 1 (1976)

M.F. Greaves (London), Cell Surface Receptors: A Biological Perspective

F. Macfarlane Burnet (Melbourne), The Evolution of Receptors and Recognition in the Immune System

K. Resch (Heidelberg), Membrane Associated Events in Lymphocyte Activation

K.N. Brown (London), Specificity in Host-Parasite Interaction

Volume 2 (1976)

D. Givol (Jerusalem), A Structural Basis for Molecular Recognition: The Antibody Case

B.D. Gomperts (London), Calcium and Cell Activation

M.A.B. de Sousa (New York), Cell Traffic

D. Lewis (London), Incompatibility in Flowering Plants

A. Levitski (Jerusalem), Catecholamine Receptors

Volume 3 (1977)

J. Lindstrom (Salk, California), Antibodies to Receptors for Acetylcholine and other Hormones

M. Crandall (Kentucky), Mating-Type Interactions in Micro-organisms

H. Furthmayr (New Haven), Erythrocyte Membrane Proteins

M. Silverman (Toronto), Specificity of Membrane Transport

Volume 4 (1977)

M. Sonenberg and A.S. Schneider (New York), Hormone Action at the Plasma Membrane: Biophysical Approaches

H. Metzger (NIH, Bethesda), The Cellular Receptor for IgE

T.P. Stossel (Boston), Endocytosis

A. Meager (Warwick) and R.C. Hughes (London), Virus Receptors

M.E. Eldefrawi and A.T. Eldefrawi (Baltimore), Acetylcholine Receptors

Volume 5 (1978)

P.A. Lehmann F. (Mexico), Stereoselective Molecular Recognition in Biology

A.G. Lee (Southampton), Fluorescence and NMR Studies of Membranes

L.D. Kohn (Bethesda, Maryland) Relationships in the Structure and Function of Receptors for Glycoprotein Hormones, Bacterial Toxins and Interferon

Series B

Published

The Specificity and Action of Animal, Bacterial and Plant Toxins (B1)
edited by P. Cuatrecasas (Burroughs Wellcome, North Carolina)

Intercellular Junctions and Synapses (B2)
edited by J. Feldman (London), N.B. Gilula (Rockefeller University, New York) and J.D. Pitts (University of Glasgow)

Microbial Interactions (B3)
edited by J.L. Reissig (Long Island University, New York)

Specificity of Embryological Interactions (B4)
edited by D. Garrod (University of Southampton)

In preparation

Taxis and Behavior (B5)
edited by G. L. Hazelbauer (Wallenberg Laboratory, Uppsala)

Virus Receptors (B6)
edited by L. Philipson, L. Randall (University of Uppsala) and K. Lonberg-Holm (Du Pont, Delaware)

Receptors and Recognition

Receptors

Edited by
P. Cuatrecasas
*Wellcome Research Laboratory,
Research Triangle Park, North Carolina*
and
M. F. Greaves
*ICRF Tumour Immunology Unit,
University College London*

and Recognition

Series A

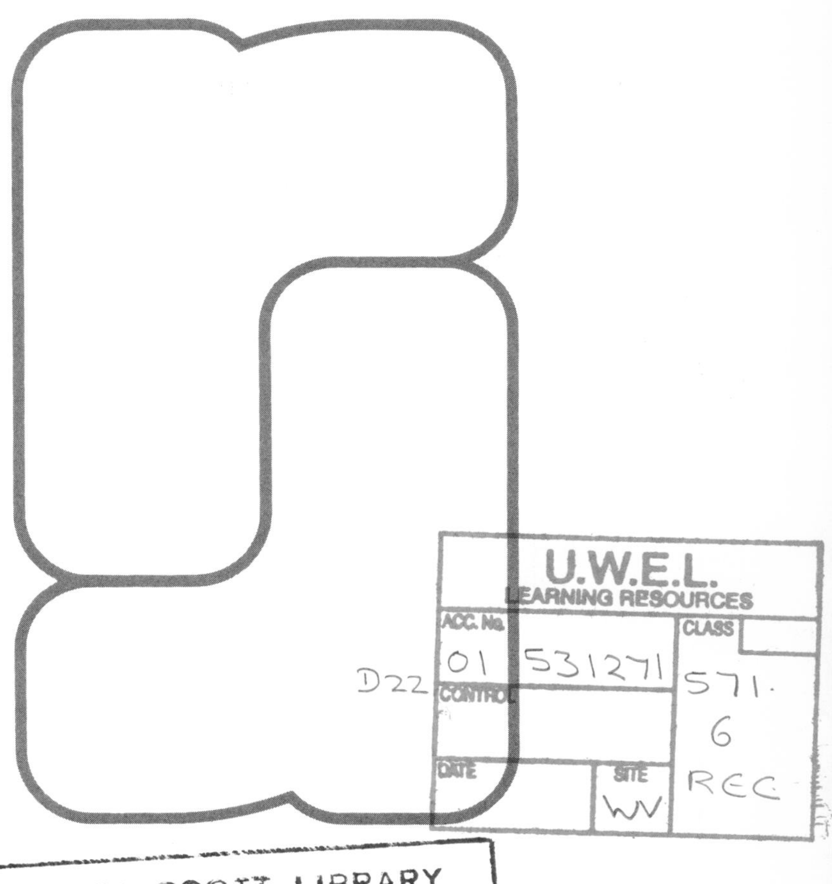

LONDON
CHAPMAN AND HALL

A Halsted Press Book
John Wiley & Sons, New York

First published 1978
by Chapman and Hall Ltd
11 New Fetter Lane, London EC4P 4EE

Typeset by C. Josée Utteridge-Faivre and printed in Great Britain by Cambridge University Press

ISBN 0 412 15290 8 (cased edition)
ISBN 0 412 15280 0 (paperback edition)

Distributed in U.S.A. by Halsted Press,
a Division of John Wiley & Sons, Inc., New York

Library of Congress Cataloging in Publication Data (Revised)

Main entry under title:

Receptors and recognition.

Includes bibliographical references.
1. Cellular recognition. 2. Binding sites
(Biochemistry) I. Cuatrecasas, P. II. Greaves, Melvyn F.
QR182.R4 574.8'76 75-44163
ISBN 0-470-26464-0

Contents

Preface

As in previous volumes of this series of 'Receptors and Recognition', Volume 6 seeks to demonstrate and promote the value of comparing and integrating recognition events in widely different biological systems in the search for common mechanisms or evolutionary links. The topics presented in this volume illustrate the realization and potential of cross-fertilization of methodologies and concepts.

The chapter by Eytan and Kanner describes imaginatively the current state of the art in the exciting area of membrane reconstitution. The methodology for incorporating isolated or purified membrane proteins into artificial lipid model systems while retaining or regaining the original function of that protein (ie. reconstitution) is becoming an indispensable tool in membrane biochemistry. This approach can provide a functional assay for membrane proteins, an especially important tool for proteins (eg., promoters of solute transfer proteins, hormone receptors, etc.) whose functionality is lost or disturbed upon solubilization. In addition, it is possible by this approach to examine systematically the effects of changing the membrane milieu (ie. membrane composition) on the function of a given protein. Such studies are also providing insights into the normal mode of membrane assembly *in vivo.*

O'Brien's chapter is a lucid and critical review of one of the most important model systems in membrane biology, that of the light-sensitive glycoprotein, rhodopsin. Like the erythrocyte, which has taught us profoundly important lessons about oxygen transport and the chemistry of heme proteins, the study of rhodopsin illustrates how broad can be the research opportunities of an apparently narrow, specialized problem. O'Brien describes vividly how photoreceptor outer segments may be the most convenient preparation of excitable membranes and provide a nearly ideal system for study. Apart from providing insights into the molecular events of vision research, the outer segments are an excellent model for studying the role of phospholipids in the permeability of membranes. Rhodopsin is a remarkably interesting protein – it is a lipoprotein, a glycoprotein, and a conjugated protein with a chromophore that absorbs in the visible portion of the spectrum, and it is an integral membrane protein that spans the membrane and may even itself be an ion pore under conditions of illumination.

The chapter by Fain represents a most perceptive and imaginative exposition of the broad field of hormones, membranes and cyclic nucleotides. Perhaps the most crucial questions and problems in this field have been identified and addressed in a fashion that brings together devergent facts and separates fancy. A number of extremely important and exciting topics are raised in a critical and interesting manner. Among those are included the nature of hormone receptors, the possible mechanisms for receptor-adenylate cyclase recognition and coupling, agonist-specific desensitization and tachyphylaxie and the possible roles of calcium, adenosine and phosphotidylinositol breakdown. The complexities and highly integrated nature of interactions in these systems are brought out forcefully, and an interesting case (the activation of glycogen phosphorylase in rat liver by hormones and cyclic nucleotides) is presented which illustrates these points convincingly. Despite the enormous problems and difficulties of these biological systems, progress into the molecular bases of some fundamental phenomena are now occurring rapidly!

May 1978

P. Cuatrecasas
M.F. Greaves

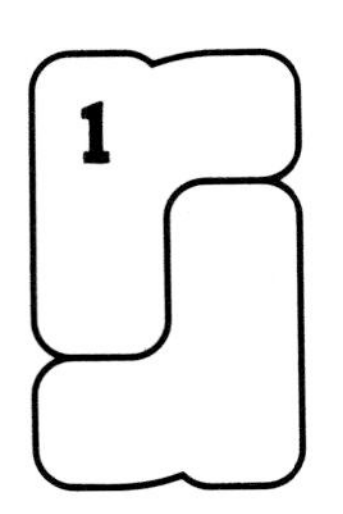

Hormones, Membranes and Cyclic Nucleotides

JOHN N. FAIN
Division of Biology and Medicine,
Brown University,
Providence, R.I. 02912, U.S.A.

Abbreviations

Cyclic AMP	Adenosine 3′, 5′-monophosphate
Cyclic GMP	Guanosine 3′, 5′-monophosphate
EGTA	Ethylene glycol bis (β-aminoethylether)–N, N-tetraacetic acid
GppNHp	Guanyl-5′-yl imidodiphosphate
CDP diglyceride	Cytidine diphosphate diglyceride

Acknowledgements

This review was prepared while I was on sabbatical leave as Macy Faculty Scholar and Visiting Fellow of Clare Hall in the Department of Zoology at Cambridge University. The research from my laboratory was supported by a research grant from the National Institute of Arthritis, Metabolism and Digestive Diseases (AM10149).

Receptors and Recognition, Series A, Volume 6
Edited by P. Cuatrecasas and M.F. Greaves
Published in 1978 by Chapman and Hall, 11 New Fetter Lane, London EC4P 4EE

1.1 INTRODUCTION

Many hormones interact with receptors on the outer surface of the plasma membrane of cells. The hormone–receptor complex alters the enzymatic activity of membrane-bound enzymes. Adenylate cyclase is probably the best example of a membrane-bound enzyme which is activated by hormones and other agents. Adenylate cyclase catalyses the formation of cyclic AMP which serves as an intracellular messenger to activate cytoplasmic enzymes.

Cyclic AMP was discovered twenty years ago by Sutherland and Rall (Rall and Sutherland, 1958; Sutherland and Rall, 1958). They found that the activation of glycogen phosphorylase in slices of dog and cat liver by catecholamines and glucagon involved an unknown heat-stable compound which turned out to be cyclic AMP. Sutherland and Rall were able to obtain cell-free liver homogenates in which activation of phosphorylase could be readily demonstrated upon the addition of hormones. The response of the liver homogenate occurred in two steps. In the first step, a particulate fraction of the liver homogenates (which contained fragments of the plasma membrane) produced a heat-stable factor (cyclic AMP) in the presence of glucagon or catecholamines. In the second step, the addition of cyclic AMP activated phosphorylase in the supernatant fraction of the homogenate.

Most effects of hormones on intact cells which are thought to involve changes in the activity of membrane-bound enzymes have been difficult to demonstrate in cell-free systems. However, adenylate cyclase is an exception since it responds to a large array of hormones added to membrane preparations derived from a wide variety of cells.

A formidable problem in all biological research is the development of suitable assay procedures for substances present in very low concentrations. Cyclic AMP was no exception. For ten years the only available assay was the activation of dog liver phosphorylase introduced by Sutherland and Rall. Their laboratory was virtually the only one in which this assay could be reliably performed. However, the recent development of sensitive, accurate, fast and relatively inexpensive assays has resulted in a tremendous expansion of cyclic nucleotide research. The radioligand binding assays using protein kinase introduced by Gilman (1970) and antibodies against cyclic AMP by Steiner, Parker and Kipnis (1972) are

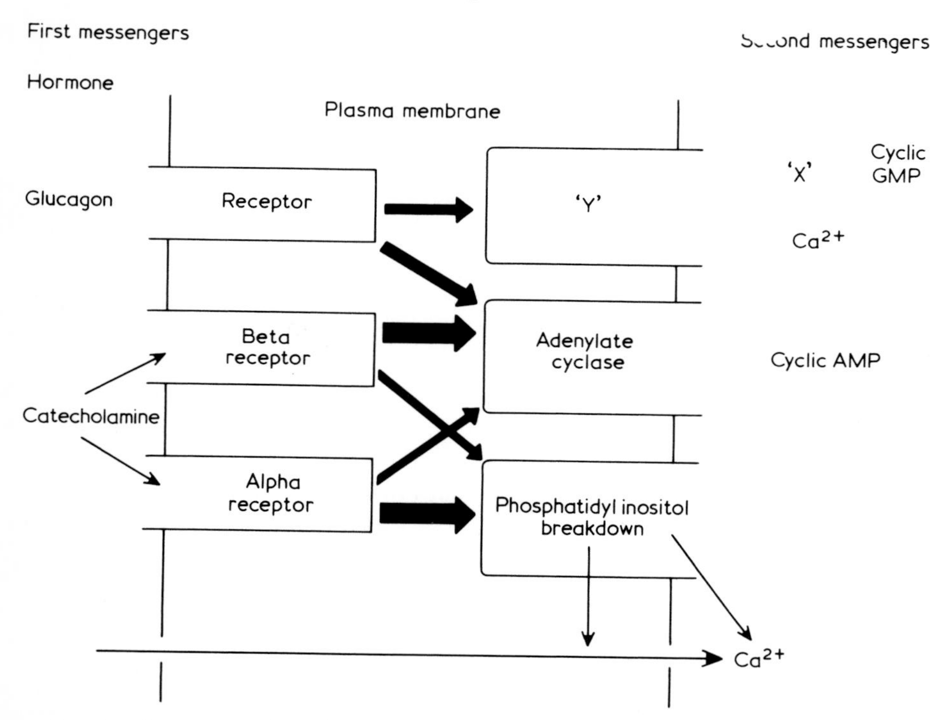

Fig. 1.1 Representation of the interaction of hormones (first messengers) with the cell membrane and the release of intracellular signals (second messengers). Cyclic AMP accumulation is accelerated by the interaction of glucagon or β-catecholamines with receptors fixed on the outer surface of the plasma membrane. The hormone–receptor complex rapidly activates adenylate cyclase located on the inner surface of the plasma membrane. There are receptors in the membrane for α-catecholamines which alter an unknown enzyme 'Y' resulting in an increase in the intracellular accumulation of 'X'. The two leading candidates for 'X' are Ca^{2+} and cyclic GMP. Possibly 'X' is intracellular Ca^{2+} which is responsible for the increase in cyclic GMP. The model postulates that the glucagon–receptor complex is able to activate other enzymes besides adenylate cyclase such as 'Y'. The difference between α and β effects of catecholamines is that β predominantly activate adenylate cyclase while α activate 'Y'. The breakdown of phosphatidylinositol may be an example of the 'Y' activation by α adrenergic agonists which is related in some unknown fashion to the release of 'trigger' Ca^{2+} and the entry of extracellular Ca^{2+}.

now universally used and have supplanted all other assays for cyclic nucleotides. The antibody procedure can be used to detect other cyclic nucleotides such as cyclic GMP by using the proper antisera.

The original concept of Sutherland and his associates (Robison *et al.*, 1971) was that hormones are first messengers which carry information to the plasma membrane of cells where they interact with receptors (Fig. 1.1). The receptors are localized in the plasma membrane and do not carry the hormone into the cytosol or nucleus of the cell. If the hormone–receptor complex does not enter the cytosol then a mechanism is required to transfer information into the cell. Possibly some hormones regulate cellular function by activating membrane-bound proteins such as those involved in the regulation of hexose and Ca^{2+} transport. However, in many cases, there is a need for a second messenger to transfer information to the cell's internal machinery. Robison *et al.*, (1971) suggested that cyclic AMP was not the only second messenger. At the moment cyclic GMP and Ca^{2+} are the leading candidates to join cyclic AMP as second messengers.

There has been an unfortunate tendency to assume that if a hormone or other agent alters the level of a cyclic nucleotide then all of the effects of that hormone are secondary to alterations in cyclic AMP or GMP. There are fads and fashions in science as in all other aspects of human endeavour. At the moment, cyclic nucleotides are in style and we are being deluged with reports on measurements of cyclic nucleotides in every possible system. Investigators have reported correlations between cyclic nucleotides and almost every known physiological and pharmacological response. However, data which do not agree with cyclic nucleotide involvement in a given response have often been ignored in the past.

A prior article in this series by Sonenberg and Schneider (1977) emphasized biophysical approaches to hormone action at the plasma membrane. Cuatrecasas and Hollenberg (1976) have reviewed the role of membrane receptors in hormone action. I have reviewed the role of cyclic nucleotides in the hormonal regulation of fat cell metabolism (Fain, 1973a, 1977, 1978). The present chapter emphasizes the effects of hormones on the coupling of the hormone–receptor complex to adenylate cyclase in the plasma membranes of mammalian cells. Particular attention is given to the hypothesis that activation of adenylate cyclase is not the sole effect of hormones such as glucagon, catecholamines and thyrotropin.

1.2 CRITERIA USED FOR INVOLVEMENT OF CYCLIC AMP IN HORMONE ACTION

The original criteria established by Sutherland and his associates (Robison *et al.*, 1971) for involving cyclic AMP in a given effect of a hormone were as follows:

(1) Adenylate cyclase activity of broken cell preparations should be stimulated by the hormone. Hormones which do not give the particular response should be without effect on adenylate cyclase.

(2) Intracellular cyclic AMP should be elevated by concentrations of the hormone which are capable of producing the physiological response. The log-dose response curve for the physiological response and for elevation of cyclic AMP should be identical. Furthermore the elevation in cyclic AMP should precede the physiological response rather than follow it. Hormones which do not produce the given response should also be inactive with respect to elevating cyclic AMP.

(3) The response to the hormone should be potentiated by inhibitors of cyclic AMP phosphodiesterase.

(4) Exogenous cyclic AMP should mimic the action of the hormone.

In addition there are now several other criteria which can be added:

(5) There should be protein kinase activity of extracts from the particular cell which is activated by cyclic AMP.

(6) The addition of cholera toxin should mimic the effect of the hormone if cyclic AMP is elevated by the toxin.

(7) If the enzyme responsible for the response to the hormone (lipase in the case of lipolysis or phosphorylase for glycogenolysis) is known it should be activated by the addition of cyclic AMP and protein kinase.

The effects of many hormones meet these criteria and can be explained by activation of adenylate cyclase. However, it is possible that some hormones exert effects independent of cyclic AMP. The concentration of hormone required to give a half-maximal activation of cyclic AMP accumulation is often far greater than that required to give a half-maximal response to the hormone. Few studies have shown a good correlation between levels of cyclic AMP and the physiological response under a wide variety of conditions. Most reports presented as proof of this criteria measured cyclic AMP in the presence of methyl xanthines and the response to the hormone in the absence of methyl xanthine.

The lack of correlation between cyclic AMP and the given response may sometimes be more apparent than real if there is a high basal

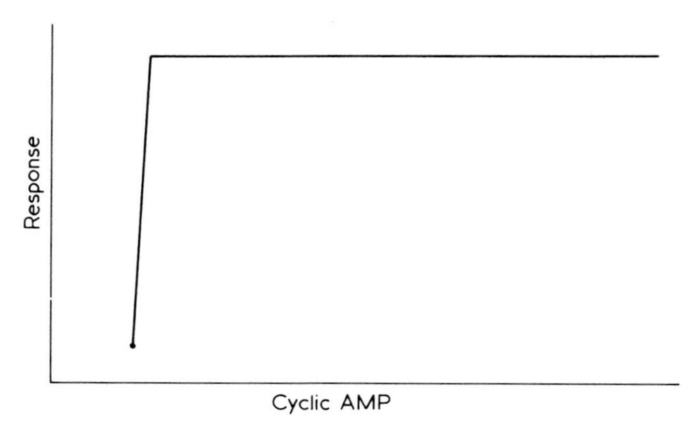

Fig. 1.2 The theoretical relationship between total cyclic AMP accumulation and a particular biological response when the pool of cyclic AMP is a small part of the basal pool. This model assumes that doubling the physiologically important pool of cyclic AMP is sufficient to give a maximal response. The large excess capacity for cyclic AMP accumulation may have other functions unrelated to the short-term response.

concentration of cyclic AMP. Basal cyclic AMP values may represent the total cyclic AMP content of several different cell types and intracellular compartments. If the active cyclic AMP pool is a small fraction of the total, it could be doubled without any detectable elevation in total cyclic AMP. This may be the explanation for the almost all-or-nothing effect of hormones on cyclic AMP illustrated in Fig. 1.2 which is frequently seen in many cells. The accumulation of large amounts of cyclic AMP usually requires an unphysiologically high concentration of hormone and the presence of methyl xanthine. Neither are characteristic of *in vivo* conditions. A concentration of hormone sufficient to give a maximum biological response *in vivo* seldom elevates cyclic AMP by more than a factor of two.

Another possibility is that there are other second messengers formed in the presence of hormones which potentiate the action of cyclic AMP. This hypothesis suggests that cyclic AMP is not the primary signal, but rather is responsible for a prolonged and sustained response to hormones.

In rat liver cells, an increase in cytosol Ca^{2+} activates glycogen phosphorylase. An attractive hypothesis is that low concentrations of hormones increase the Ca^{2+} pool in contact with glycogen phosphorylase while higher concentrations elevate cyclic AMP. The action of Ca^{2+} may be more transient than that of cyclic AMP which activates protein kinases that phosphorylate regulatory enzymes. Possibly Ca^{2+} or other unknown messengers might increase the sensitivity of protein kinase to cyclic AMP by binding to and inactivating inhibitors of protein kinase.

1.3 ADENYLATE CYCLASE

This review is largely concerned with the regulation of adenylate cyclase which is located in the plasma membrane of animal cells. Adenylate cyclase catalyzes the conversion of a molecule of ATP to cyclic AMP and pyrophosphate. Mg^{2+} is required for the reaction and the substrate is probably a Mg^{2+}–ATP complex.

Under physiological conditions, the formation of cyclic AMP is virtually irreversible due to the very high level of pyrophosphatase. The formation of cyclic AMP utilizes the equivalent of two high energy phosphate bonds per molecule of cyclic AMP which is formed. The phosphodiester bond of cyclic AMP is a high energy bond since the free energy of hydrolysis is about 12 kcal mol^{-1}. This does not appear to have any particular role in cyclic AMP activation of protein kinase. Rather it may be involved in ensuring that the conversion of cyclic AMP to 5′-AMP by cyclic AMP phosphodiesterase is an irreversible reaction.

Adenylate cyclase is widely distributed throughout the animal kingdom and is found in all nucleated cells (Robison *et al.*, 1971). The location of adenylate cyclase in the plasma membrane was first demonstrated by Davoren and Sutherland (1963). In the early studies of Sutherland on adenylate cyclase, the tissues were homogenized under conditions in which the plasma membrane was present as large pieces which sedimented with nuclei in the 600 *g* precipitate. Subsequent studies have clearly demonstrated the presence of adenylate cyclase in the plasma membrane (Perkins, 1973). The plasma membrane appears to be the only location of this enzyme in mammalian cells.

Adenylate cyclase is probably located on the inner surface of the plasma membrane. The addition of ATP to intact cells results in its cleavage by membrane-bound ATPase but no formation of cyclic AMP occurs (Robison *et al.*, 1971). Treatment of whole cells with proteolytic enzymes reduced the activity of ATPase but not that of adenylate cyclase (Oye and Sutherland, 1966).

The isolation and purification of adenylate cyclase from membranes has proven to be a difficult problem. Cyclase activity can be solubilized rather readily with detergents. However, in most studies solubilization results in an irreversible loss of hormone sensitivity. With few exceptions, it has been difficult to obtain reconstitution of hormone-sensitive adenylate cyclase by mixing fractions containing catalytic activity with inactive fractions containing hormone bound to a receptor.

Levey (1971, 1973) solubilized adenylate cyclase activity in cat heart homogenates with the non-ionic detergent Lubrol–PX. The solubilized activity had a molecular weight in the range of 100 000–200 000. The solubilized preparation was adsorbed to DEAE-cellulase and separated from most of the detergent by sequential elution with neutral buffers of increasing ionic strength. The addition of highly purified bovine phosphatidyl inositol restored catecholamine responsiveness to the solubilized enzyme. The response to glucagon was not restored by phosphatidylinositol. However, high concentrations of phosphatidylserine (75 times more phospholipid than was required for restoration of catecholamine sensitivity) restored glucagon sensitivity. Unfortunately as yet, no other investigators have been able to restore hormone sensitivity by specific phospholipids. Neer (1973) solubilized renal medullary adenylate cyclase with the same detergent (Lubrol-PX). The solubilized enzyme was no longer responsive to hromones but hormonal sensitivity could be restored by removal of the detergent.

1.3.1 Hormone receptors and adenylate cyclase

The activation of adenylate cyclase results from the interaction of hormones with receptors on the outer surface of the plasma membrane of cells (Fig. 1.3). Sutherland and his associates (Robison *et al.*, 1971) originally suggested that adenylate cyclase is a large protein which spans the plasma membrane and contains binding sites for hormones on the outer surface and catalytic sites for cyclic AMP formation on the inner surface.

The evidence for fixed receptors on the outer surface of the plasma membrane is as follows: Schimmer *et al.* (1968) covalently linked ACTH to cellulose producing a large molecular weight entity which is believed to be too large to permeate cells. The ACTH linked to cellulose was able to stimulate cyclic AMP-mediated steroidogenesis in cultured adrenal cells. Rodbell *et al.* (1970) found that trypsin treatment of intact fat cells abolished the activation of adenylate cyclase by glucagon. Neither basal adenylate cyclase nor fluoride-stimulated cyclase was affected by this treatment. These studies suggested that the glucagon receptor is a protein or linked in some way to trypsin-sensitive proteins. Further support has come from studies in which the addition of specific anti-sera abruptly

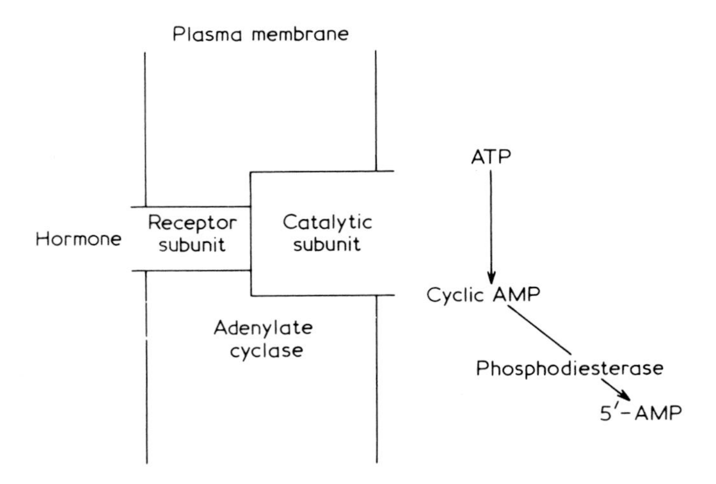

Fig. 1.3 Original concept of adenylate cyclase as containing the receptor site for hormones.

terminated the action of peptide hormones *in vitro* (Pastan *et al.*, 1966). However, the best evidence for the location of hormone receptors on the exterior part of the plasma membrane came from the extensive studies with labelled hormones in which direct binding to cells or plasma membrane preparations was assayed (Catt and Dufau (1977).

The simple hypothesis that adenylate cyclase is a large protein which spans the plasma membrane and contains a binding site for hormones as well as the catalytic site (Fig. 1.3) is not very probable. Birnbaumer and Rodbell (1969) first demonstrated that at least five different hormones activate adenylate cyclase activity in fat cell ghosts. It is unlikely that a single protein (i.e. adenylate cyclase) could contain five distinct receptor sites. Fig. 1.4 depicts the activation of the same adenylate cyclase by many different hormones each binding to distinct receptors. The most likely hypothesis is that there are a number of receptors clustered around each adenylate cyclase molecule.

A convincing demonstration that the hormone receptors are discrete units quite different from adenylate cyclase has come from the work of Schramm and his associates (Orly and Schramm, 1976; Schramm *et al.*, 1977). They were able to fuse turkey erythrocytes containing catecholamine receptors with mouse erythroleukemia Friend cells (F-cells). The F-cells have no detectable β-adrenergic receptors but do have adenylate cyclase activity. The catalytic activity of adenylate cyclase in turkey

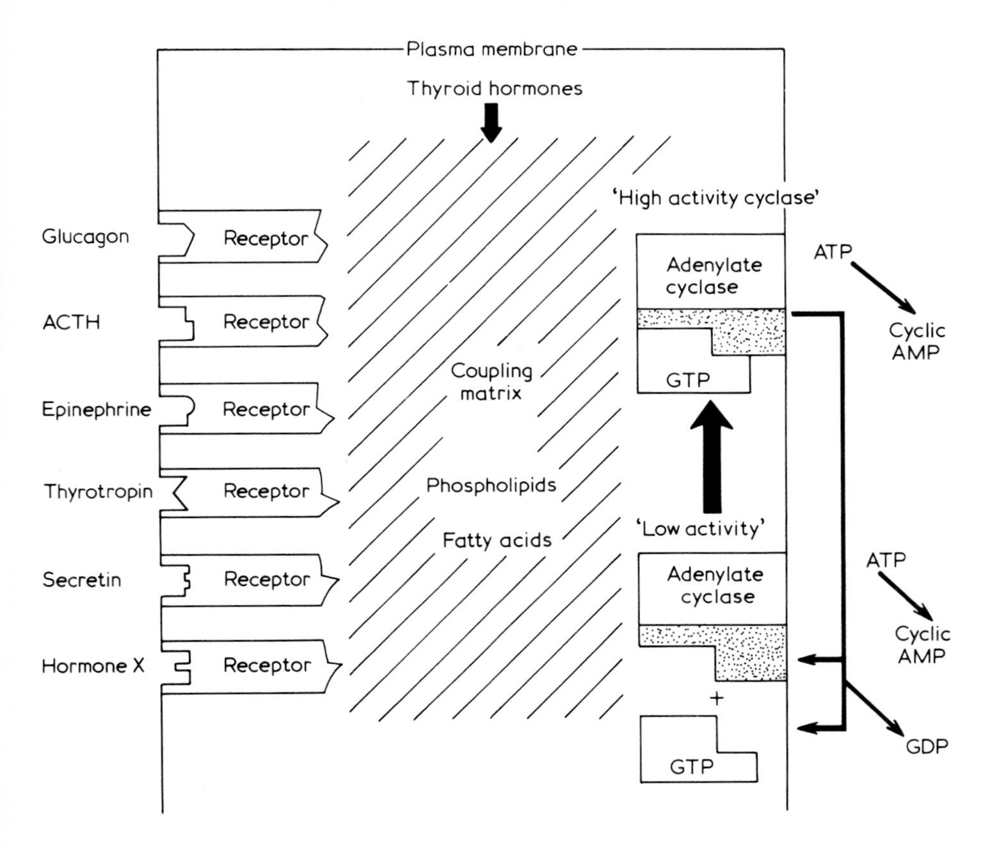

F ig. 1.4 Role of guanine nucleotides in activation of adenylate cyclase by hormones. The model has the following features.

1. A single adenylate cyclase activated by a variety of hormones.
2. A coupling process involving phospholipids which is regulated by thyroid hormone (T–3).
3. Adenylate cyclase activity is on a protein separate from the receptors.
4. An obligatory GTP requirement for adenylate cyclase activation.
5. Activation may involve reaction of inactive or low-activity adenylate cyclase with GTP and another protein in the presence of the hormone–receptor complex. Whether the high activity cyclase has a larger or smaller molecular weight remains to be established. I prefer the view that dissociation of the catalytic subunit from a regulatory subunit (represented by a stippled box in the model) after binding to a GTP binding protein is the more probable explanation for activation of adenylate cyclase.

erythrocytes was inactivated by *N*-ethylmaleimide or heat treatment. Within a few minutes after fusion of turkey erythrocytes containing catecholamine receptors with F-cells containing adenylate cyclase, a marked activation of the enzyme by catecholamines was detected. These data demonstrate that the hormone receptors have a common attachment site by which they can activate adenylate cyclase in foreign cells.

Insel *et al.* 1976 have suggested that hormone receptors and adenylate cyclase are the products of separate genes. It is unlikely that two different genes would be involved in the formation of single protein. Limbird and Lefkowitz (1977) found that, after solubilization of frog red cell membranes, it was possible to separate adenylate cyclase activity from catecholamine binding. The conclusion of all the studies reported to date is that the hormone receptors and adenylate cyclase are separate proteins coded by different genes.

If hormone receptors are in molar excess with respect to adenylate cyclase in the plasma membrane a mechanism is needed by which the hormone–receptor complex comes in contact with adenylate cyclase. Cuatrecasas (1974) and De Haen (1976) proposed that hormone receptors are not physically associated with adenylate cyclase in the free state. Cuatrecasas (1974) described this hypothesis initially as a two-step fluidity hypothesis for activation of adenylate cyclase by hormones. Subsequently, Cuatrecasas *et al.* (1975) called it the mobile receptor hypothesis. However, the original designation of the hypothesis as a two-step fluidity model for activation of adenylate cyclase by hormones may be more likely. If the receptor sites for hormones are not a part of adenylate cyclase and there is an excess of receptors over cyclase molecules then any hypothesis for activation of cyclase involves at least two steps.

The plasma membrane is now viewed as consisting of proteins floating in a mobile lipid bilayer (Singer and Nicolson, 1972). Lateral mobility of hormone–receptor complexes is easy to envisage in such a membrane structure. In most cells, the entire process of hormone binding to receptor and activation of adenylate cyclase occurs without any appreciable lag period (10 seconds or less). Thus, whatever lateral movement of the hormone–receptor complex is required for activation of adenylate cyclase occurs fairly rapidly.

1.3.2 Coupling of hormone–receptor complexes to adenylate cyclase

The most important discovery in recent years with regard to the activation of adenylate cyclase was the recognition that guanyl nucleotides are involved in hormonal activation of this enzyme. Rodbell *et al.* (1971)

found that guanyl nucleotides were required for the activation of hepatic adenylate cyclase by glucagon. Subsequent studies using guanyl 5′-yl imidodiphosphate(GppNHp) have shown that this analog of GTP activates adenylate cyclase and potentiates the response to hormones in a wide variety of cells (Londos *et al.*, 1974; Pfeuffer and Helmreich, 1975). The GppNHp analog is probably more potent than GTP because its terminal phosphate cannot participate in phosphotransferase (GTPase) reactions which are quite active in membrane preparations. At the high concentrations of commercially available preparations of ATP (1 mM) used in most adenylate cyclase assays, there are often enough guanine nucleotides present as contaminants to permit activation of adenylate cyclase by hormones (Kimura and Nagata, 1977).

Pfeuffer (1977) found that over 95% of the GTP binding sites in pigeon erythrocytes could be separated from adenylate cyclase by sucrose density gradient centrifugation. A small amount of GTP binding to a protein of M_r = 42 000 remained which was associated with adenylate cyclase. If this protein was removed from detergent-solublized preparations of adenylate cyclase by the use of a novel GTP-Sepharose derivative there was a decrease in the ability of GTP and fluoride to activate adenylate cyclase. Reactivation of adenylate cyclase occurred when the two fractions (the adenylate cyclase after passage through GTP-Sepharose columns and the guanyl nucleotide binding protein eluted from the GTP-Sepharose column) were recombined. Guanyl-nucleotide binding proteins isolated from pigeon erythrocyte membranes reactivated rabbit myocardial adenylate cyclase preparations depleted of binding proteins. Pfeuffer (1977) suggested that catecholamine activation of adenylate cyclase is mediated through the guanyl nucleotide binding protein.

Activation of adenylate cyclase by guanine nucleotides may occur by a mechanism involving dissociation of an inactive cyclase in the presence of GTP plus binding protein to give an active catalytic subunit. Possibly (as shown in Fig. 1.4), the GTP interaction with binding protein is facilitated by the hormone–receptor complex. The binding protein in the presence of bound GTP combines with a protein subunit of adenylate cyclase resulting in dissociation of the catalytic subunit from the regulatory subunit which binds the GTP regulatory protein. Alternatively, the complex formed between adenylate cyclase and the GTP binding protein may be the active (high activity) state of adenylate cyclase. Fig. 1.4 does not distinguish between these possibilities.

A third mechanism for activation of adenylate cyclase might involve binding of the hormone–receptor complex to low-activity adenylate cyclase which is converted to a high-activity form by GTP in the presence of a guanyl nucleotide binding protein. Either model depicts activation of cyclase as involving at least three steps and as many different proteins. This may explain the great difficulty most investigators have had in purifying adenylate cyclase.

We do not yet know how GTP activates adenylate cyclase. The model shown in Fig. 1.4 suggests that activation of adenylate cyclase is due to binding of GTP to a protein which forms a complex with adenylate cyclase converting it from a low to high activity state.

Rendell *et al.* (1977) have suggested that the adenylate cyclase system oscillates between states of low and high activity through a process which is regulated by the rate of GTP hydrolysis at the guanine–nucleotide binding site. This is an extension of the three-state kinetic model for the activation of adenylate cyclase which incorporates GTPase as an essential part of the model. The model suggests that the stimulation of adenylate cyclase by GppNHp results from the inability of GTPase to cleave this nucleotide which results in retention of adenylate cyclase in the high activity state.

(a) *Modulation of adenylate cyclase by lipids*

Engelhard *et al.* (1976) found that the fatty acids of plasma membrane phospholipids from mouse LM cells could be altered by culturing these cells in the process of linoleate. In control cells there was no detectable linoleate in membrane phospholipids. If cells were cultured in the presence of linoleate this fatty acid accounted for 23% of the fatty acid content of membrane phospholipids. The basal adenylate cyclase activity was doubled and the response to prostaglandin E_1 was tripled as a result of prior growth of cells in the presence of linoleate. There was no correlation between membrane fluidity (at least as based on viscosity measurements) and adenylate cyclase activity in either the presence or absence of prostaglandin E_1 (Engelhard *et al.*, 1976).

Arachidonic acid accounts for 25–35% of the total fatty acid content of phospholipids from liver membranes of rats fed a diet containing 5% oil (Brivio-Haugland *et al.*, 1976). If rats were fed on diets deficient in essential fatty acids there was a marked drop in the arachidonic acid content of all phospholipids, particularly phosphatidylcholine and inositol (Brivio-Haugland *et al.*, 1976). The basal adenylate cyclase activity of liver membranes from these rats was reduced by 50% (Brivio-Haugland *et al.*, 1976). However the percentage increases in adenylate cyclase due to glucagon, fluoride or glucagon in the membranes from rats fed on a diet deficient in essential fatty acids were normal. There was a slight decrease in 5′-nucleotidase and an increase in ATPase activity of membranes from the EFA-deficient animals. It is disappointing that the drastic changes in fatty acid composition had so little effect on adenylate cyclase.

Another approach to the role of lipids in adenylate cyclase regulation was that of Shier *et al.* (1976). They found that the addition of lysolecithin or Triton X-100 to membrane preparations from 3T3 mouse fibroblasts inhibited adenylate cyclase activity but activated guanylate cyclase. Whether lysolecithin has any physiological role in cyclase regulation remains to be established but is especially intriguing.

(b) *Role of thyroid hormones as regulators of coupling between hormone–receptor complexes and adenylate cyclase*

Thyroid hormones may potentiate catecholamine action by regulating the number of hormone receptors or the interaction of the catecholamine–receptor complexes with adenylate cyclase. Treatment with large amounts of thyroid hormones increased the number of catecholamine receptors in the heart (Ciaraldi *et al.*, 1977; Williams *et al.*, 1977). However, there was no significant decrease in catecholamine binding of membrane from hypothyroid rats (Ciaraldi *et al.*, 1977). These data suggest that there is something special about the hyperthyroid state which increases the number of catecholamine receptors in the heart.

In fat cells from hypothyroid or hyperthyroid rats the number of catecholamine binding sites were unaltered (Table 1.1). Maximal cyclic AMP accumulation was markedly depressed in fat cells from hypothyroid rats although neither the number, affinity or character of catecholamine binding sites nor the amount of fluoride-stimulatable adenylate cyclase was altered by hypothyroidism (Malbon *et al.*, 1978).

In view of the role of GTP in modulating hormone activation of adenylate cyclase we attempted to restore the sensitivity of fat cell ghosts from hypothyroid rats with GppNHp (Malbon *et al.*, 1978). Fat cell ghosts were isolated and then incubated for one hour at 4°C with GppNHp. However, what resulted was an even greater difference in response between ghosts from normal and hypothyroid rats. Malbon *et al.* (1978) postulated that thyroid hormones may affect lipolysis by altering the coupling of the hormone–receptor complex to adenylate cyclase.

Glucocorticoids are another example of hormones which regulate coupling of hormonal–receptor complexes to adenylate cyclase

Table 1.1 Effect of hypothyroidism on hormone binding and cyclic AMP metabolism of fat cells. The data are taken from the report of Malbon *et al.* (1978a). Hypothyroid rats were prepared by giving rats an iodine-deficient diet for 21–24 days along with propylthiouracil in the drinking water. Binding sites for (–) [^{3}H] dihydroalprenolol were measured at a concentration of 1 nM using fat cell membranes. Cyclic AMP accumulation and glycerol release were measured after 2 and 25 minutes incubation respectively of isolated cells in the presence of 100 μm epinephrine. Adenylate cyclase activity was measured using fat cell ghosts pre-incubated for 1 h in an ice bath with or without 50 μM GppNHp. The ghosts were washed and then incubated for 20 min with 100 μM norepinephrine or 10 mM sodium fluoride.

	Binding sites for 1 nM dihydroalprenolol (fmol/mg protein)	Cyclic AMP accumulation (pmol/10^6 cells)	Glycerol release (μmol/10^6 cells)	Adenylate cyclase activity (nmol/mg protein)		
				+ GppNHp	GppNHp + norepinephrine	Fluoride
Hypothyroid	16	<5	8.4	0.4	3.5	5.9
Control	20	100	9.7	0.9	8.8	6.1

(Rajerison *et al.*, 1974). The number and affinity of vasopressin receptors in rat kidney medulla membranes is unaffected by adrenalectomy or glucocorticoids. However, the ability of vasopressin to activate adenylate cyclase in these membranes was reduced by adrenalectomy and enhanced by glucocorticoid treatment (Rajerison *et al.*, 1974). It may be significant that both thyroid hormones and glucocorticoids primarily act as permissive agents which potentiate the action of other hormones such as catecholamines but have little effect on their own.

(c) ***Adenylate cyclase uncoupled from hormone receptors***

Haga *et al.* (1977) isolated a novel variant of S49 mouse lymphoma cells possessing β-adrenergic receptors and adenylate cyclase activity which was insensitive to catecholamines. They termed these clones 'uncoupled' since it appeared that hormone binding had been uncoupled from activation of adenylate cyclase. The number of β-adrenergic binding sites was unaltered. However, neither catecholamines nor prostaglandin E_1 affected cyclic AMP accumulation by the clones but the cells showed a normal rise in cyclic AMP to cholera toxin. Haga *et al.* (1977) suggested that the defect in these clones was in a factor unrelated to hormone binding to receptors which was required for the activation of adenylate cyclase.

Ross and Gilman (1977) found that adenylate cyclase activity in detergent-solubilized lymphoma membranes was readily inactivated by incubation at 37°C for 20 min. Adenylate cyclase activity was restored by incubating with detergent-solubilized membranes from clones of lymphoma cells which do not have adenylate cyclase activity. Their results suggested that a protein is present in the clones which is able to replace the readily inactivated protein involved in activation of adenylate cyclase. This protein may be the guanyl nucleotide binding protein of Pfeuffer (1977).

Henneberry *et al.* (1977) have shown that the addition of butyrate to cultured HeLa cells resulted in an increased activation of adenylate cyclase by catecholamines. The adenylate cyclase activity of these cells was not altered by butyrate but rather the coupling of the hormone–receptor complex to adenylate cyclase and the number of catecholamine receptors. Low concentrations of butyrate induced a large increase in the number of catecholamine receptors but there was little activation of adenylate cyclase. Higher concentrations of butyrate were required to couple the hormone–receptor complexes to adenylate cyclase. At the appropriate concentration of butyrate, catecholamine receptors and adenylate cyclase are present but there is little activation of cyclase.

The mechanisms by which low concentrations of butyrate increase the number of catecholamine receptors while higher concentrations act as coupling agents is not known. However, butyrate inhibits the growth of HeLa cells and increases the content of the sialyl transferase which forms the glycosphingolipid GM_3 (Ginsberg *et al.*, 1973).

1.3.3 Agonist-specific desensitization

Perkins *et al.* (1975) reported a specific loss of sensitivity to catecholamines by human astrocytoma cells incubated with norepinephrine for two hours. Although the cells had been washed free of catecholamine following the 120 min incubation, they failed to respond to the addition of fresh norepinephrine. However they did respond normally to prostaglandin E_1 following this incubation with catecholamine. Since the response to prostaglandin E_1 was normal, the activation of cyclic AMP phosphodiesterase appeared an unlikely explanation.

In the rat pineal gland (Kebabian *et al.*, 1975) or the frog erythrocyte (Mukherjee *et al.*, 1975) prior exposure to catecholamines desensitized the cells to a second addition of catecholamines. In the pineal gland, the reduction in catecholamine receptors appeared to be independent of any changes in protein synthesis (Kebabian *et al.*, 1975).

Desensitization appeared to result from a reduction in the number of catecholamine binding sites after prior exposure to these hormones (Kebabian *et al.*, 1975; Mukherjee *et al.*, 1975). In frog erythrocyte membranes, the inactive catecholamine binding sites could be recovered by exposure of the membrane to GppNHp or GTP (Mukherjee and Lefkowitz, 1976). The rapid recovery of catecholamine binding sites suggested that agonist-specific desensitization results from the formation of hormone–receptor complexes which are unable to activate adenylate cyclase. These complexes dissociate very slowly except in the presence of guanyl nucleotides. The formation of these inactive complexes is a reasonable explanation for the so-called agonist-specific desensitization. It may be more accurate to refer to this as agonist-specific inactivation of receptors. Catecholamines antagonists appear to bind to the same receptor site as catecholamines. The difference is that the antagonist–receptor complex does not result in the formation of inactive receptors even after prolonged exposure.

1.3.4 Feedback regulation of adenylate cyclase

A general shortcoming in cyclic nucleotide research has been the failure to realise that cellular responses to hormones operate as closed-loop control systems. Fig. 1.5 indicates that responses of cells to hormonal

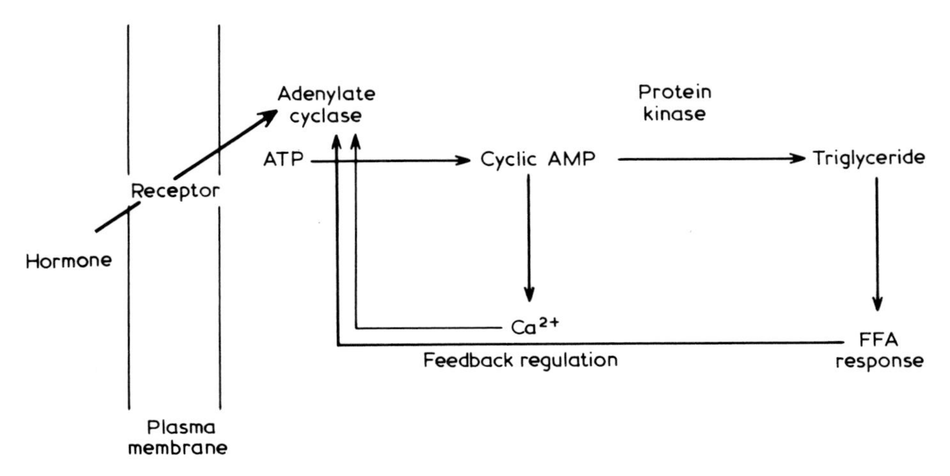

Fig. 1.5 The cyclic AMP information transfer system as closed-loop control involving feedback regulation of adenylate cyclase. The response chosen to illustrate feedback regulation is lipolysis by rat fat cells. It remains to be demonstrated whether the fatty acid inhibition of adenylate cyclase activation by hormones is on hormone binding or activation of cyclase by the hormone–receptor complex. The feedback regulation of adenylate cyclase by Ca^{2+} is intended to point out only that changes in the level of Ca^{2+} may influence cyclic AMP and that the cytosol Ca^{2+} concentration can be influenced by cyclic AMP.

stimuli are dependent not only on present and past stimuli but also the response itself. Ordinarily, as in Fig. 1.4, the formation and action of cyclic AMP is depicted as an open-loop control system. It remained for Rasmussen and Goodman (1977) to point out that the cyclic AMP system is no exception to the paradigm that the physiological response influences the stimulus. In other words, there is a feedback relationship between stimulus and response. An interesting aspect of the cyclic AMP system is that both cyclic AMP and the response elicited in a particular cell may act as feedback regulators. Recent work has been concerned less with feedback regulation than with so-called agonist specific receptor inactivation which is limited to a particular agonist. In contrast, feedback regulation will reduce the response of adenylate cyclase to all stimulators.

(a) *Feedback regulation by fatty acids*

The incubation of rat fat cells with lipolytic agents results in the release of antagonists (feedback regulators) to the medium which reduce the cyclic AMP elevation seen in response to a second addition of hormone (Ho and Sutherland, 1971; Manganiello *et al.,* 1971). The antagonists which accumulate in the medium in response to epinephrine are not hormone-specific since they block the subsequent response to ACTH and glucagon (Ho and Sutherland, 1971; Manganiello *et al.,* 1971). In contrast to agonist-specific receptor inactivation, the effects of feedback regulators on fat cell adenylate cyclase are readily reversed by washing the cells and incubating them in fresh buffer. The feedback regulators do not affect cyclic AMP phosphodiesterase activity of fat cells (Pawlson *et al.,* 1974). There is general agreement that the feedback regulators released during incubation of fat cells with lipolytic agents are non-dialyzable substances which bind to medium albumin (Fain and Shepherd, 1975; Ho and Sutherland, 1971).

The feedback regulator released to the medium is probably not adenosine. Fain and Shepherd (1975) suggested that free fatty acids were the feedback regulators of lipolysis. It is not the fatty acid content which is important but rather the molar ratio of free fatty acids to albumin in the medium. Free fatty acids are insoluble in ordinary buffer and will only accumulate in the medium if serum or albumin is present to bind the fatty acids. Rodbell (1965) originally reported that lipolysis by incubated fat cells stops after enough fatty acids accumulate to saturate the primary binding sites on medium albumin. There are two to three tight binding sites for fatty acids on each mole of albumin. Scow (1965) found that, in perfused adipose tissue, lipolysis was inhibited at molar ratios of free fatty acid to albumin above two.

Fain and Shepherd (1975) postulated that the free fatty acids released during lipolysis act as product inhibitors of triglyceride lipase and feedback regulators of adenylate cyclase at molar ratios of medium free fatty acid to albumin above two. There was a direct inhibitory effect of fatty acids on adenylate cyclase which could be demonstrated using isolated rat fat cell ghosts.

There are species differences in regulation of adenylate cyclase since cyclic AMP accumulation in chicken fat cells does not appear to be subject to feedback regulation. This was reflected in the insensitivity of chicken fat cell adenylate cyclase to direct inhibition by free fatty acids (Malgieri *et al.*, 1975).

The feedback regulation of rat fat cell adenylate cyclase by free fatty acids is a good example of closed-loop control (Fig. 1.5). It is reasonable that the free fatty acids derived from triglyceride lipolysis should inhibit both adenylate cyclase and triglyceride lipase when the primary binding sites on albumin are saturated. Otherwise there would be an intracellular accumulation of free fatty acids which have deleterious effects on fat cell energy metabolism (Angel *et al.*, 1971).

Ho *et al.* (1975) have suggested that feedback regulation involves more than the free fatty acid to albumin ratio. The best possibility is that an endoperoxide derived from the further metabolism of arachidonic acid released during lipolysis acts as a regulator of adenylate cyclase.

It is unlikely that prostaglandins have much of a role in short-term regulation of fat cell adenylate cyclase. Illiano and Cuatrecasas (1971) claimed that indomethacin affected cyclic AMP accumulation and lipolysis, but others found no affect of this drug on fat cell metabolism during short-term incubations (Dalton and Hope, 1973; Fain *et al.*, 1973; Fredholm and Hedqvist, 1975). Indomethacin is a potent inhibitor of the cyclooxygenase which is responsible for the first step in the conversion of arachidonic acid to endoperoxides, thromboxanes and prostaglandins (Ferreira and Vane, 1974; Hamberg and Samuelsson, 1974; Hong and Levine, 1976). Dalton and Hope (1974) found that indomethacin was an effective inhibitor of prostaglandin synthesis.

Recently, Shepherd and Fain (unpublished results, 1978) found some evidence that regulation of prostaglandin synthesis might be involved in the lipolytic action of growth hormone and glucocorticoids. There is a one to two hour lag before lipolysis is accelerated by growth hormone or glucocorticoids (Fain *et al.*, 1971; Fain and Saperstein, 1970). The lag period for activation of lipolysis by these hormones, unlike that for cholera toxin, appears to involve RNA and protein synthesis. Incubation of fat cells with growth hormone and sometimes with dexamethasone for three to four hours resulted in an increase in the subsequent ability of catecholamine to increase cyclic AMP. Prior incubation of fat cells for four hours with either growth hormone or glucocorticoids enhanced the subsequent activation of adenylate cyclase in fat cell ghosts by GppNHp and fluoride. Incubation of fat cells for four hours prior to isolation of ghosts with 10 μg ml^{-1} of indomethacin mimicked the effects of growth hormone and glucocorticoids on adenylate cyclase. These results are in contrast to our prior negative results with respect to indomethacin effects on lipolysis and cyclic AMP accumulation in intact cells but, in those studies, the length of incubation did not exceed two hours.

Possibly it takes some time before the effects of prostaglandins or thromboxanes on fat cells are apparent. Growth hormone and glucocorticoids may accelerate synthesis of a protein which is able to block cyclooxygenase or some further step in the metabolism of arachidonic acid. The fact that glucocorticoids are well-known anti-inflammatory agents like aspirin and indomethacin suggests a common link through inhibition of prostaglandin formation.

Whether it is prostaglandins of the E series or prostaglandin endoperoxides which inhibit rat fat cell adenylate cyclase is not known. Gorman (1975) found that the direct addition of PGH_2 (15-hydroxy-9-peroxidoprosta-5, 13-dienoic acid) at concentrations in the range of 0.3–30 μM to fat cell ghosts resulted in a reduction in adenylate cyclase activity. The real question is whether PGH_2 occurs naturally in fat cells and if its formation is accelerated during lipolysis.

1.3.5 Ca^{2+} as a regulator of adenylate cyclase

Rasmussen (1970) and Berridge (1975) have emphasized the possible importance of the interactions between cyclic AMP and calcium in hormone action. Berridge (1975) has reviewed the evidence that an important component in regulation of adenylate cyclase is the interaction between calcium and cyclic AMP.

The relationships between cyclic AMP and calcium were divided into two categories by Berridge (1975). Monodirectional systems are those in which cyclic AMP enhances the calcium signal. Bidirectional systems are those in which the effect of cyclic AMP is opposed by calcium.

Bidirectional control was originally described by Goldberg *et al.* (1974) as the Yin Yang or dualism hypothesis of biological control. Goldberg *et al.* (1974) had cyclic GMP rather than calcium as the signal interacting with cyclic AMP. In systems exhibiting Yin Yang behaviour (bidirectional control) one type of stimulus increases cyclic AMP and decreases cyclic GMP while the reverse is true for the other type of stimulus. However, Goldberg *et al.* (1974) also recognized that in some cells both cyclic AMP and GMP are elevated at the same time.

The role of cyclic GMP in hormone action remains to be established. Several hormones stimulate its formation but no direct effect of any hormone on guanylate cyclase has yet been established. Furthermore there is no known biological process which is regulated by cyclic GMP.

The elevation of cyclic GMP seen after addition of hormones is generally dependent on extracellular calcium. Guanylate cyclase appears to be a soluble enzyme which is activated by elevations in intracellular calcium (Goldberg and Haddox, 1977). There is a good correlation between the effect of hormones on extracellular calcium uptake and cyclic GMP. A most useful hypothesis at the moment is to consider elevations in cytosol cyclic GMP as reflections of a rise in cytosol calcium at the sites of guanylate cyclase activity.

Probably cyclic GMP, cyclic AMP and calcium are connected to each other by an intricate maze of feedback relationships. It is important to sort out which second messenger is stimulated directly by a hormonal signal, which is elevated indirectly and which functions as a feedback regulator of the primary signal (Rasmussen and Goodman, (1977).

Berridge and Rapp (1977) suggested that a very important aspect of feedback regulation is the ability of cyclic AMP to modulate calcium homeostasis (Fig. 1.6). In a cyclic AMP-calcium closed-loop control system a rise in cyclic AMP increases cytosol calcium and calcium inhibits adenylate cyclase (Berridge and Rapp, 1977; Rapp and Berridge, 1977). The same type of control results if cyclic AMP decreases cytosol calcium but calcium is now an activator of adenylate cyclase (Fig. 1.6).

Brooker (1973) has shown oscillations in cyclic AMP content of frog heart and a periodic release of cyclic AMP by slime moulds has been found (Gerisch and Wick, 1975). Possibly oscillations in calcium and cyclic AMP drive many cellular pacemakers and biological rhythms (Berridge and Rapp, 1977; Rapp and Berridge, 1977). Neuronal oscillations, myogenic oscillations in smooth muscle and pacemaker oscillations in cardiac muscle have all been attributed to the rhythmic change in calcium and cyclic AMP. Probably in these systems calcium is the primary signal whose level is affected by cyclic AMP. However, in the closed loop system involving cyclic AMP and calcium (Fig. 1.6) regulation of either signal will affect the level of the other signal.

A calcium-regulated protein modulator has been described by Brostrom *et al.* (1975) which activates brain adenylate cyclase. The protein modulator was originally discovered as a calcium-dependent factor which activated cyclic GMP phosphodiesterase in brain (Cheung, 1971; Kakiuchi and Yamazaki, 1970. This phosphodiesterase may also account for cyclic AMP hydrolysis in some tissues (Wang, 1977). Since the protein modulator increases cyclic AMP and decreases cyclic GMP it would primarily be of value in bidirectional control systems as a feedback signal.

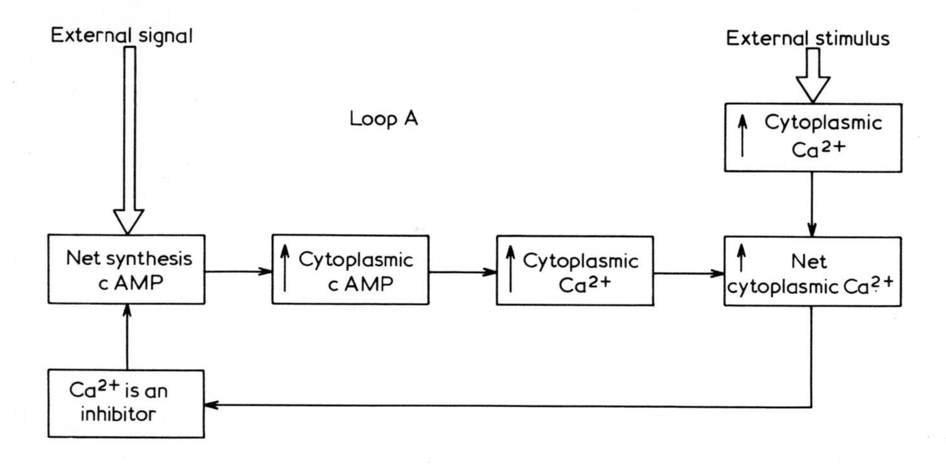

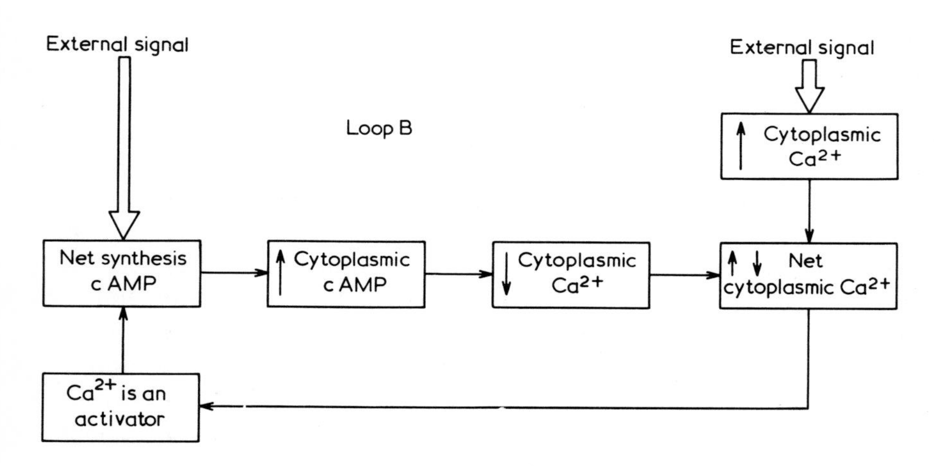

Fig. 1.6 A summary of cyclic AMP (cAMP) and Ca^{2+} feedback interactions arranged as closed-loop feedback control. The figure indicates that the concentrations of cyclic AMP and Ca^{2+} are influenced by feedback loops within cells as well as external signals. In loop A, an increase in the intracellular level of cAMP will lead to an increase in the level of Ca^{2+} in, for example, β-cells from pancreatic islets, liver, and insect and mammalian salivary glands. In loop B, an increase in cAMP will lower the level of Ca^{2+} and will thus oppose the action of external signals which act to increase cytoplasmic Ca^{2+} – such antagonistic effects are found in lymphocytes, mast cells and smooth muscle. The figure is reproduced, with permission, from the report of Berridge and Rapp (1977).

A major problem in discussing calcium regulation of adenylate cyclase is the possibility that in some tissues low concentrations of calcium activate while high concentrations of calcium inhibit adenylate cyclase (Rasmussen and Goodman, 1977). Calcium is known to inhibit adenylate cyclase from rat heart (Drummond and Duncan, 1970), renal cortex (Streeto, 1969) and liver (Pointer *et al.*, 1976).

1.3.6 Adenosine and prostaglandins as regulators of adenylate cyclase

Adenosine and prostaglandins of the E series are unique regulators of adenylate cyclase. In some tissues adenylate cyclase is stimulated (Sattin and Rall, 1970; Skolnick and Daly, 1977) while in other tissues adenylate cyclase is inhibited by prostaglandins E_1 or E_2 (Fain and Shepherd, 1978; Ismail *et al.*, 1977; Trost and Stock, 1977). It has been difficult to demonstrate a direct inhibition of adenylate cyclase by adenosine or prostaglandins.

There are multiple effects of prostaglandins of the E series on incubated pieces of rat adipose tissue. Prostaglandin E, paradoxically, stimulated cyclic AMP accumulation and inhibited lipolysis in rat adipose tissue (Butcher and Baird, 1968). Butcher and Baird found that prostaglandins increased cyclic AMP in the non-fat cells of adipose tissue (primarily the endothelial cells of the extensive capillary network in adipose tissue). In isolated fat cells prepared by collagenase digestion of adipose tissue the rise in cyclic AMP due to catecholamines was inhibited as was lipolysis by prostaglandin (Butcher and Baird, 1968). These results demonstrate the problems one can encounter in cyclic nucleotide research unless the studies are done with a single cell type or broken cell preparation derived from these cells.

Adenosine was first found to be a potent stimulator of cyclic AMP accumulation in brain slices by Sattin and Rall. A unique difference between adenosine and other stimulators of brain cyclic AMP accumulation was the finding that theophylline blocks the rise in cyclic AMP due to adenosine (Sattin and Rall, 1970). Antagonism has also been seen in cells where adenosine lowers cyclic AMP (Fain *et al.*, 1978).

The possibility that adenosine is a direct neurotransmitter in the central nervous system seems plausible. Adenosine potentiates the rise in cyclic AMP due to α-adrenergic amines in the central nervous system. However, there is little potentiation of the β response to catecholamines by adenosine (Skolnick and Daly, 1977).

Fain *et al.* (1972) found that adenosine was a potent inhibitor of cyclic AMP accumulation and lipolysis by rat fat cells. Schwabe *et al.* (1973) found that incubated fat cells released adenosine to the medium. There is no evidence to date that adenosine accumulation in the medium is stimulated by lipolytic hormones under physiological conditions (Fain and Shepherd, 1978; Fain *et al.*, 1978; Schwabe *et al.*, 1973). Fat cells are ordinarily incubated in medium containing bovine fraction V albumin to bind fatty acids. Most of the available preparations contain enough adenosine deaminase activity as a contaminant to deaminate any free adenosine which accumulates in the medium (Fain and Shepherd, 1978).

Some adenosine appears to be bound to the fat cells since lipolysis and cyclic AMP accumulation by rat fat cells are accelerated by the addition of large amounts of exogenous adenosine deaminase to the medium (Fain, 1973; Fain and Weiser, 1975; Schwabe and Ebert, 1972). In our experience, using albumin with some adenosine deaminase activity, maximal accumulation of adenosine is seen at the start of the incubation (Fain and Shepherd, 1978; Fain *et al.*, 1978). Any further adenosine released by the cells appears to be deaminated to inosine. A net accumulation of adenosine is only seen in the presence of inhibitors of adenosine deaminase and high concentrations of lipolytic agents or valinomycin. The rise in adenosine release correlated with the rise in fat cell AMP (Fain and Shepherd, 1978). Valinomycin in an ionophore which increases potassium flux across biological membranes and also inhibits lipolysis (Fain and Loken, 1971). The inhibition of lipolysis is probably secondary to the marked drop in ATP and rise in AMP seen after addition of this ionophore to fat cells (Fain and Loken, 1971).

The mechanism by which adenosine restrains cyclic AMP accumulation is not yet established. Fain *et al.* (1972) found an inhibition of adenylate cyclase activity of membrane preparation derived from fat cells by high concentration of adenosine and similar results have been seen by others (Fain *et al.*, 1978; Trost and Stock, 1977). However, inhibition of the adenylate cyclase activity of broken cell preparations due to high concentrations of adenosine may occur by a different mechanism than is observed in intact fat cells.

The inhibition of adenylate cyclase activity of fat cell membranes by high concentrations of adenosine may be at the 2′, 5′-dideoxyadenosine site. In broken cell preparations this analog of adenosine, which cannot be deaminated or phosphorylated, is a better inhibitor of adenylate cyclase than is N^6-phenylispropyl adenosine (Fain *et al.*, 1972; 1978; Trost and Stock, 1977). The phenylisopropyl analog is also resistant to deamination. In intact fat cells the reverse is true for 2′, 5′-dideoxy-adenosine is less effective than phenylisopropyl adenosine as an inhibitor of cyclic AMP accumulation (Fain *et al.,* 1972). Table 1.2 summarizes the differences between these two adenosine analogs. An interesting feature is the lack of antagonism between methylxanthines and 2′, 5′-dideoxyadenosine (Fain *et al.,* 1978). This lends further support to the hypothesis that 2′, 5′-dideoxyadenosine interacts at a different site than does adenosine.

In isolated liver cells, 2′, 5′-dideoxyadenosine is an inhibitor of cyclic AMP accumulation (Fain and Shepherd, 1977) and of adenylate cyclase (Fain and Shepherd, 1977; Londos and Preston, 1977). While 2′,5′-dideoxy-adenosine markedly reduced the elevation in cyclic AMP seen with glucagon it did not affect the rise in glucose release. Similar results have been seen in fat cells when 2′, 5′-dideoxyadenosine reduced the rise in cyclic AMP due to hormones without affecting lipolysis (Fain *et al.,* 1978). One possibility is that adenylate cyclase activity in intact cells is compartmentalized. Perhaps most of the detectable cyclic AMP accumulation results from activation by high levels of hormone of an adenylate cyclase pool which is unrelated to physiological responses but is inhibited by dideoxyadenosine. In contrast, adenosine may inhibit a small pool of adenylate cyclase which generates the cyclic AMP involved in lipolysis. In membrane preparations the site, for adenosine inhibition of adenylate cyclase is either lost or masked by a large number of physiologically unimportant adenylate cyclase molecules.

Alternatively, the hormone–receptor complex may activate some membrane event which regulates lipolysis and the stimulation of adenylate cyclase is only an amplification signal. This hypothesis suggests that adenosine inhibits the ability of the hormone–receptor complex to affect all membrane-bound enzymes while 2′, 5′-dideoxyadenosine inhibits adenylate cyclase activation by the hormone–receptor complex.

The available evidence suggests that adenosine binds to a site on the plasma membrane of fat cells at which it inhibits adenylate cyclase in intact cells. Adenosine covalently linked to stachyose had been found to be equipotent with adenosine as an inhibitor of cyclic AMP accumulation by rat and chicken fat cells. The adenosine-stachyose derivative is thought to be too big to enter fat cells. Since there was no lag before cyclic AMP accumulation was inhibited by adenosine linked to stachyose and it was equipotent with adenosine, it seems unlikely that it was cleaved to adenosine (Fain and Shepherd, 1978). Olsson *et al.* (1976) originally demonstrated that this large molecular weight complex of adenosine

Table 1.2 Ineffectiveness of 2′, 5′-dideoxyadenosine as an inhibitor of fat cell lipolysis. The data are taken from the report of Fain *et al.* (1978, Fain and Shepherd, 1978) and that of Trost and Stock (1977)

Agent	Concentration required for 50% inhibition of catecholamine stimulation of			Effect on cyclic AMP reversed by methyl xanthines
	Cyclic AMP accumulation	Lipolysis	Adenylate cyclase	
2′, 5′-dideoxyadenosine	10 μM	No effect at concentrations up to 50 μM	0.5 μM	No
Adenosine	0.01 μM	0.10 μM	10 μM	Yes
N^6 (phenylisopropyl) adenosine	0.003 μM	0.01 μM	No effect at concentrations up to 50 μM	Yes

linked to stachyose, like adenosine, was a coronary vasodilator in dogs.

The inhibition by adenosine of adenylate cyclase activity in broken cell preparations has been difficult to demonstrate. It is not known whether it is the ability of adenosine to bind to specific receptor sites which is lost or the ability of the adenosine once bound to these receptors to inhibit adenylate cyclase. Malbon *et al.* (1978) were able to demonstrate relatively high affinity binding sites for adenosine in isolated fat cells plasma membrane preparations with an apparent dissociation constant of 9.5 μM. It was only possible to detect adenosine binding in the presence of a potent inhibitor of adenosine deaminase such as deoxycoformycin or erythro-9(2-hydroxy-3-nonyl)adenine. The number of high affinity binding sites for adenosine on fat cell plasma membrane preparations was similar to that for β-adrenergic binding sites but there was no competition for binding between these ligands (Malbon *et al.*, 1978). The binding of adenosine was inhibited by as little as 0.1–1.0 μM theophylline which indicates that adenosine binding is far more sensitive to theophylline than is inhibition of cyclic AMP phosphodesterase (Malbon *et al.*, 1978).

Possibly the marked potentiation by methyl xanthines of cyclic AMP accumulation in many cells is due to antagonism of endogenous adenosine restraint of adenylate cyclase. Sattin and Rall (1970) first observed in brain slices that methyl xanthines blocked the rise in cyclic AMP due to adenosine. Ordinarily, the elevation of cyclic AMP due to hormones is potentiated by methyl xanthines which are thought to inhibit cyclic AMP phosphodiesterase (Robison *et al.*, 1971). However, other inhibitors of this enzyme such as papaverine (Clark *et al.*, 1974; Schwabe *et al.*, 1972) actually potentiated the rise in cyclic AMP due to adenosine in human astrocytoma cells (Clark *et al.*, 1974).

A large part of the rise in cyclic AMP due to methyl xanthines may be due to reversal of an inhibitory constraint on adenylate cyclase by endogenous adenosine. In the absence of adenosine there is no effect of methyl xanthines on cyclic AMP content of fat cells. Two examples are known. Schwabe and Ebert (1972) found that if a small number of fat cells were incubated with isoproterenol there was a large accumulation of cyclic AMP which was not potentiated by theophylline. If more cells were present there was little increase in cyclic AMP due to isoproterenol, but theophylline now markedly potentiated cyclic AMP accumulation. The second example is chicken fat cells where methyl xanthines did not affect the rise in cyclic AMP due to glucagon (Fain and Shepherd, 1978). Chicken fat cells did respond to methyl xanthines in the presence of added adenosine. There appeared to be little endogenous adenosine present because of a high adenosine deaminase activity of chicken fat cells (Fain and Shepherd, 1978).

1.3.7 Cholera toxin and guanyl nucleotides as regulators of adenylate cyclase

Cholera appears to result from an irreversible activation of adenylate cyclase in the small intestine which markedly increases fluid secretion (Van Heyningen, 1977). Cholera toxin can activate adenylate cyclase in any mammalian cell which contains this enzyme and receptors for cholera toxin. The receptor for cholera toxin (Holmgren *et al.*, 1975; Van Heyningen *et al.*, 1971) is monosialo ganglioside GM_1 (galactosyl-*N*-acetyl galactosaminyl-(sialyl)-galactosyl-glucosylceramide).

The cholera toxin molecule is composed of two non-identical subunits in the ratio of A_1 B_5 (Gill and King, 1975). The B subunit is involved in the binding of toxin to the ganglioside GM_1 and has a formula weight of 11 604. The primary sequence of amino acids has been determined (Kurosky *et al.*, 1977; Lai, 1977) for the B chain and resembles the B subunits of thyrotropin and the gonadotropins.

The A subunit does not bind to the ganglioside GM_1 but rather is involved in the activation of adenylate cyclase (Gill and King, 1975). The A subunit alone can activate adenylate cyclase in broken cell preparations from pigeon erythryocytes without any lag period in the presence of NAD (Gill and King, 1975; Van Heyningen and King, 1975). In intact cells, the A subunit is inactive and there is an appreciable delay before adenylate cyclase is activated by cholera toxin. This 30 to 90 minute lag does not involve protein synthesis (Gill and King, 1975). The lag period may be required for transfer of the A subunit of cholera toxin through the intact plasma membrane or dissociation of the A subunit after binding of the B subunit to the ganglioside GM_1.

The activation of adenylate cyclase by cholera toxin requires NAD and is irreversible (Gill and King, 1975; van Heyningen, 1973; van Heyningen *et al.*, 1971). In contrast, activation of adenylate cyclase in intact cells by hormones is rapid, doesn't require NAD and is readily reversed. The NAD requirement is interesting because tetanus toxin utilizes NAD to ribosylate a protein which regulates protein synthesis. Tetanus toxin action is also virtually irreversible and involves binding to a different ganglioside (Van Heyningen, 1973). The role of NAD in cholera toxin action remains to be elucidated. There is evidence that cholera toxin has

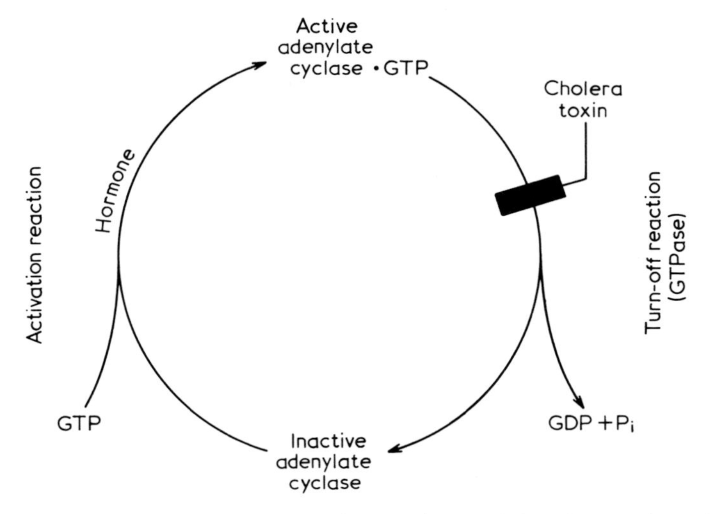

Fig. 1.7 Model of regulation of adenylate cyclase activity. Conversion of the adenylate cyclase from an inactive to an active state requires hormone–induced introduction of GTP to the regulatory site. The turn-off reaction occurs by hydrolysis of the bound GTP. Continuous hydrolysis of GTP depends on concurrent operation of both the activation and the turn-off reactions, and is therefore hormone-dependent. Cholera toxin inhibits the turn-off reaction and has no effect of the activation reaction. The figure is reproduced, with permission from the report of Cassell and Selinger (1977).

NAD glycohydrolase activity (Moss *et al.*, 1976).

The most interesting advance in our understanding of the mode of cholera toxin action involves guanine nucleotides Flores and Sharp (1975) pointed out that another irreversible activator of adenylate cyclase was guanyl-5′-yl imidiodiphosphate (GppNHp). In liver homogenates, the addition of cholera toxin blocked activation of adenylate cyclase by GppNHp and vice versa (Flores and Sharp, 1975). Bennett *et al.* (1975) observed that the adenylate cyclase activity of membranes from fat cells incubated with cholera toxin was markedly enhanced by GTP. More recently, Glossman and Struck (1977) observed that, in adrenal membranes exposed to cholera toxin, GTP was as effective an activator of adenylate cyclase as GppNHp. In control membranes GTP was much less effective than GppNHp. It has been suggested that cholera toxin blocks the dissociation and breakdown of GTP bound to a site which regulates adenylate cyclase (Cassell and Selinger, 1977; Glossman and Struck, 1977; Levinson and Blume, 1977). Cassell and Selinger (1977) have proposed that continuous GTP hydrolysis at the regulatory guanyl nucleotide binding site is a mechanism for turning off adenylate cyclase. Cholera toxin is postulated to inhibit the guanosinetriphosphatase (GTPase) reaction (Fig. 1.7).

Regulation of GTP availability for binding to adenylate cyclase may be an unrecognised site for hormone action. It is known that hormones can regulate the dissociation and breakdown of GTP bound to adenylate cyclase. Cassell and Selinger (1976) have seen an elevated GTPase activity in turkey erythrocytes after incubation with catecholamines. This has not yet been observed in other cells where the rate of non-specific hydrolysis of GTP is much higher. The activation of GTPase by catecholamine would also serve as a mechanism by which elevations in cyclic AMP act as feedback inhibitors of adenylate cyclase.

Yamamura *et al.* (1977) found that the GTPase activity of fat cell membrane preparations is much more sensitive to trypsin treatment than the proteins involved in GTP activation of cyclase. Ordinarily fat cell GTPase is so active that special conditions are required to demonstrate any activation of adenylate cyclase by GTP. Whether this represents GTPase activity associated with adenylate cyclase or some other activity remains to be demonstrated.

1.4 CYCLIC GMP

There has been a lot of excitement about cyclic GMP since George *et al.* reported (1970) that acetylcholine elevated cyclic GMP in the heart. A wide variety of hormones and other agents have been found in elevate cyclic GMP in many different cells. Goldberg and Haddox (1977) have summarized the known regulators of cyclic GMP in a recent review.

The cause and effect relationship between cyclic GMP levels and the metabolism of cells remains to be elucidated. It is possible to see elevations in cyclic GMP in a given cell either before or co-incident with the physiological response to the hormone. However, it has been difficult to mimic the metabolic effects of hormones which elevate cyclic GMP by the addition of cyclic GMP or its analogs.

Most of the recent research on the mode of action of cyclic GMP on cellular metabolism is based on the assumption that it works like cyclic AMP. Since the only clearly established action of cyclic AMP in animal cells is the activation of protein kinases a search has been made for cyclic GMP-dependent protein kinases. Enzymes have been found which respond to cyclic GMP with an increase in protein phosphorylation but their function remains to be established. Activation of protein kinase specific to cyclic GMP does not involve dissociation of a regulatory from a catalytic subunit as is true of cyclic AMP-dependent protein kinases (Goldberg and Haddox, 1977).

Cyclic GMP is probably a real intracellular messenger but unfortunately we have not yet found out what it does. The original hope that the role of cyclic GMP would be solved as readily as that of cyclic AMP has proven illusory. Furthermore in some cells such as fat and liver cells a bewildering array of agents can elevate cyclic GMP.

The metabolic effects of insulin on both liver and fat cells are generally opposite to those of agents that elevate cyclic AMP. It has been an attractive hypothesis to assume that cyclic GMP might be elevated by insulin and oppose cyclic AMP. Insulin does elevate GMP in fat and liver cells. (Fain and Butcher, 1976; Illiano *et al.*, 1973; Pointer *et al.*, 1975). However, cyclic GMP is elevated by cholinergic agents which do not mimic any of the metabolic effects of insulin (Fain and Butcher, 1976; Pointer *et al.*, 1975). Furthermore, agents which increase cyclic AMP also increase cyclic GMP in some cells. Goldberg and Haddox (1977) have suggested that this may reflect a complex regulation by cyclic GMP involving some additional messenger such as calcium. Another possibility is that cyclic GMP mediates the action of some hormones but acts as a feedback regulator for other hormones. Rasmussen and Goodman (1977) have pointed out examples of this with regard to cyclic AMP. Earp *et al.* (1977) found that the rise in cyclic AMP due to glucagon activated the nuclear guanylate cyclase. This suggests that the rise in hepatic cyclic GMP, due to agents which elevate cyclic AMP, is compartmentalized in the nucleus. In contrast, insulin and acetylcholine may elevate only cytosol cyclic GMP. This still leaves us with a need for additional messengers since acetylcholine has little effect on while insulin is an effective inhibitor of hepatic glycogenolysis.

1.4.1 Guanylate cyclase

Guanylate cyclase appears to be predominantly a membrane-bound enzyme which is readily solubilized (Goldberg and Haddox, 1977). Alternatively, there may be a separate guanylate cyclase present in the cytosol. The properties of the soluble cyclase are different from those of the plasma membrane-bound enzyme. These differences could result from an alteration of the properties of the particulate enzyme by the detergents used to unmask activity.

Guanylate cyclase activity has been reported in isolated nuclei (Earp *et al.*, 1977; Siegel *et al.*, 1976). Earp *et al.* (1977) found that there was an increase in cyclic GMP immunofluorescence of rat liver nuclei after glucagon administration. They isolated nuclei with guanylate cyclase activity which could be distinguished from that of the soluble and plasma membrane-bound guanylate cyclase. The addition of cyclic AMP to isolated nuclei increased the level of cyclic GMP. Possibly the rise in hepatic cyclic GMP due to glucagon (Earp *et al.*, 1977; Pointer *et al.*, 1976) and β-catecholamines (Pointer *et al.*, 1976) was secondary to an increase in cyclic AMP. The role of nuclear cyclic GMP is not known, but Earp *et al.*, (1977) suggested that it regulated transport across the nuclear membrane.

While adenylate cyclase has only been found in the plasma membranes of cells, some guanylate cyclase activity is in the cytosol and nuclei. Possibly activation of guanylate cyclase by hormones involves a second messenger. In the case of glucagon it was suggested that cyclic AMP was the second messenger which activated guanylate cyclase (Earp *et al.*, 1977). It will be interesting to determine whether insulin and acetylcholine also activate nuclear guanylate cyclase. If this is the case then either the hormones activate an unknown messenger (possibly calcium) or they penetrate cells and interact with hormone receptors in the nucleus. The current dogma is that all effects of insulin are secondary to interaction with receptors on the outside of the plasma membrane and the hormone does not enter the cells. This may be incorrect and insulin could be taken up by cells and act as its own internal second messenger. Alternatively, a metabolite of insulin could be a second messenger. Nerve growth factor has been found to be a polypeptide with a great deal of structural similarity to insulin. There is evidence that the nerve growth factor interacts with plasma membrane receptors and is also internalized where it interacts with nuclei (Bradshaw and Frazier, 1977).

The effects of various agents on guanylate cyclase are quite different from those on adenylate cyclase. No hormone has been shown to activate guanylate cyclase in membrane preparation where activation of adenylate cyclase is readily demonstrated. Guanylate cyclase in membranes is activated by fatty acids, phospholipids and detergents under conditions where adenylate cyclase activity is inhibited by these agents (Goldberg and Haddox, 1977). Some guanylate cyclases are readily activated by an oxidation process which can be promoted by peroxides, free radicals and oxidants like dehydroascorbic acid (Goldberg and Haddox, 1977; Haddox *et al.*, 1976). It has been suggested that the level of cyclic GMP in cells is is a reflection of changes in the balance between intracellular oxidation and reduction of key thiol groups (Goldberg and Haddox, 1977).

The physiological regulation of guanylate cyclase by the oxidation-reduction state is unlikely. The ability of nitrosamines (De Rubertis and Craven, 1976b) and azide (De Rubertis and Craven, 1976a; Kimura *et al.*, 1975) to elevate intracellular cyclic GMP is unaffected by the absence of extracellular calcium. In contrast, all the reported increases in cyclic GMP due to hormones are dependent on the presence of extracellular calcium. This calcium requirement was first found in the rat ductus deferens (Schultz *et al.*, 1973) and since then in rat liver (Pointer *et al.*, 1976), fat cells (Fain and Butcher, 1976), slices of dog thyroid (Van Sande *et al.*, 1975) and rat renal cortex (De Rubertis and Craven, 1976c). In broken cell preparations the soluble guanylate cyclase of rat lung (Chrisman *et al.*, 1975) and liver (Kimura and Murad, 1975) was activated while the particulate activity was inhibited by Ca^{2+}. The original suggestion by Schultz *et al.* (1973) that Ca^{2+} is the primary regulator of intracellular guanylate cyclase and elevations in cyclic GMP are reflections of a rise in intracellular Ca^{2+} has been supported by subsequent studies.

The presence of extracellular Ca^{2+} has little effect on the activation of glycogen phosphorylase in rat liver cells by α-adrenergic agents (Fain *et al.*, 1978). Since extracellular Ca^{2+} is required in order for cyclic GMP to be elevated by α-adrenergic agents is rat liver cells there is no apparent relationship between α effects of catecholamines and cyclic GMP. Possibly changes in cyclic GMP are secondary to elevations of intracellular Ca^{2+} and are involved in feedback regulation. In any case, the role of cyclic GMP in liver and fat cells is obscure since omission of extracellular Ca^{2+} has little effect on hormonal regulation of hepatic glycogenolysis or adipocyte lipolysis but does abolish the increase in cyclic GMP (Fain and Butcher, 1976).

1.5 CYCLIC NUCLEOTIDE PHOSPHODIESTERASES

The inactivation of cyclic AMP and GMP involves the cleavage of the cyclic 3′, 5′-phosphodiester bond to give the corresponding 5′-nucleotide (Robison *et al.*, 1971). Butcher and Sutherland (1962) found a soluble phosphodiesterase activity in beef heart with a K_m of around 0.1 mM for cyclic AMP which was inhibited by methyl xanthines. Many investigators in the cyclic nucleotide field assume that inhibition of cyclic AMP phosphodiesterase by methyl xanthines accounts for the biological effects of coffee, tea and other beverages containing these compounds. Millimolar concentrations of methyl xanthines inhibit cyclic nucleotide phosphodiesterase and potentiate the effects of hormones on cyclic AMP accumulation in most mammalian cells. However, drugs such as papaverine are more potent inhibitors of cyclic AMP phosphodiesterase than are methyl xanthines but do not potentiate cyclic AMP accumulation in intact cells (Appleman *et al.*, 1973; Fain, 1973b).

Probably the biological effects of methyl xanthines involve more than an inhibition of cyclic AMP phosphodiesterase. This is not to discredit the hypothesis that their biological effects *in vivo* are mediated through elevation in cyclic AMP, but rather that something besides inhibition of cyclic nucleotide hydrolysis is involved.

Methyl xanthines may act as antagonists of endogenous adenosine. In neural tissues the rise in cyclic AMP due to adenosine is antagonized by methyl xanthines, while other inhibitors of cyclic nucleotide phosphodiesterase potentiate adenosine-induced elevation of cyclic AMP (Skolnick and Daly, 1977). The adenosine and methyl xanthine antagonism has previously been discussed with regard to adenylate cyclase and could account for all or part of the elevations in cyclic AMP due to methyl xanthines.

The regulation of cyclic AMP phosphodiesterase by hormones does not appear to account for the primary action of any hormone. Whether it even accounts of the effects of methyl xanthines remains an open question. Probably changes in phosphodiesterase activity due to hormones are a result of elevated levels of cyclic AMP (Armiento *et al.*, 1972; Pawlson *et al.*, 1974). This may be part of a feedback mechanism for regulation of cyclic AMP.

Most studies on cyclic nucleotide phosphodiesterase activity have been done with the soluble fraction of cell homogenates at millimolar concentrations of cyclic AMP. This measures total hydrolysis of cyclic AMP at unphysiological concentrations of substrate. Furthermore, in many tissues, multiple forms of cyclic nucleotide phosphodiesterase activity are present. In rat liver extracts there is a form which hydrolyzes cyclic GMP but not cyclic AMP (Appleman *et al.*, 1973). Another enzymatic form has a low-substrate affinity for cyclic AMP and GMP while the hydrolysis of cyclic AMP is activated by low concentrations of cyclic GMP. The third form of activity appears to be the most important with regard to physiological regulation of cyclic AMP hydrolysis since it has a high affinity for cyclic AMP, displays negatively co-operative kinetics and is inhibited by cyclic GMP. Hormones which elevate cyclic AMP also elevate cyclic GMP in liver which may result in inhibition of the breakdown of cyclic AMP at low levels of cyclic GMP, but if the concentration increases it would promote hydrolysis of cyclic AMP. Whether cyclic GMP acts as negative feedback regulator at high cyclic AMP concentrations remains to be demonstrated.

There is a calcium-dependent modulator protein which plays an important role in regulation of cyclic nucleotide levels in brain (Wang, 1977). While the calcium modulator protein was originally described as an activator of the brain cyclic GMP phosphodiesterase (Cheung, 1971) it is now known to activate adenylate cyclase. Elevations in intracellular calcium should increase cyclic AMP and lower cyclic GMP. In tissues where there is a reciprocal relationship between cyclic AMP and GMP it is reasonable for a rise in intracellular calcium to activate adenylate cyclase as part of a termination signal which restores the original balance between the cyclic nucleotides. Radmussen and Goodman (1977) have indicated that this is probably of more importance than has been previously recognised.

Loten and Sneyd (1970) found that the prior administration of insulin resulted in an increase in the low K_m particulate cyclic AMP phosphodiesterase activity of adipose tissue. The original observation has now been confirmed in many laboratories but no direct effect of insulin on broken cell preparations has been demonstrated (Czech, 1977). The effect of insulin is only seen in crude particulate preparations and further attempts at enzyme purification have resulted in loss of the insulin effect. Possibly this effect of insulin does not account of any of the known actions of insulin on fat cells. The most interesting aspect is the mechanism by which prior treatment with insulin modifies the activity of fat cell cyclic AMP phosphodiesterase. Insulin has very little effect on the levels of cyclic AMP in fat cells (Fain, 1977) and it is unlikely that any of the effects of insulin on fat cell metabolism are secondary to alterations in cyclic AMP content (at least as measured for the total cyclic AMP pool).

Fain and Butcher (1976) found that the cyclic GMP content of rat fat cells was elevated by lipolytic agents (glucagon or catecholamines), antilipolytic agents (insulin), and by the divalent cation ionophore (A23187) or cholinergic agents which do not affect fat cell lipolysis. The elevations in cyclic GMP were dependent on the presence of extracellular calcium which did not modify the biological effects of the agents. Possibly the rise in cyclic AMP phosphodiesterase due to insulin and lipolytic agents is secondary to a rise in intracellular calcium and or cyclic GMP. However, we found that the activation of cyclic AMP phosphodiesterase due to insulin was readily demonstrated in calcium-free buffer and neither A-23187 nor cholinergic agents activated fat cell phosphodiesterase (Malbon and Fain, 1978).

Further information on cyclic nucleotide phosphodiesterases can be obtained by consulting the reviews by Appleman *et al.* (1973) and Wells and Hardmann (1977). A compilation of all the drugs which affect this enzyme, mostly at quite high concentrations, has been published which suggested that selective cyclic nucleotide phosphodiesterase inhibitors might be of great value as therapeutic agents (Chasin and Harris, 1976). However, to date, there has yet to be discovered a cyclic nucleotide analog or cyclic nucleotide phosphodiesterase inhibitor which has therapeutic value. Abnormalities in cyclic AMP metabolism have been assumed to explain everything from obesity to cancer (Weiss and Hart, 1977), but most such abnormalities have been results and not causes of disease.

1.6 PROTEIN KINASES

Activation of protein kinases by cyclic AMP is the only well-established mechanism by which cyclic AMP affects the metabolism of animals cells (Langan, 1973). The substrate specificity of cyclic AMP-dependent protein kinases is rather broad since almost any protein with accessible sites for phosphorylation will serve as a suitable substrate. Kuo and Greengard (1969) originally postulated that all the varied effects of cyclic AMP were mediated through activation of protein kinase. This hypothesis does not explain the effects of cyclic AMP on bacteria where there is a specific binding protein for cyclic AMP which regulates the attachment of RNA polymerase to specific regions of DNA (De Rubertis and Craven, 1976c). In mammalian cells there may be cyclic AMP binding proteins which bind cyclic AMP and regulate the activity of enzymes besides protein kinases.

Several laboratories suggested in 1970 (Brostrom *et al.*, 1970; Gill and Craven, 1970; Kumon *et al.*, 1972; Tao *et al.*, 1970) that cyclic AMP activated protein kinase by dissociating an active catalytic subunit from an inactive complex. The inactive complex consists of a regulatory subunit which binds cyclic AMP and a catalytic subunit which is relatively inactive when bound to the regulatory subunit. Cyclic AMP binds to the regulatory subunit and this results in dissociation of the active catalytic subunit from an inactive complex. This is usually expressed in the following equation:

$$R_2 C_2 \text{ (inactive)} + \text{Cyclic AMP} \rightleftharpoons 2 \text{ (Cyclic AMP}-\text{R)} + 2\text{C (active)}$$

Regulatory or binding subunit = R; Catalytic subunit = C

The cyclic AMP-dependent protein kinases can be divided into the so-called Type I kinases which are eluted by low salt from DEAE-cellulose columns, are easily dissociated by histones, reassociate slowly, and do not undergo self-phosphorylation. The Type II kinases are eluted from the same columns with higher salt concentrations, dissociate slowly upon addition of histone, reassociate rapidly and undergo self-phosphorylation (Rosen *et al.*, 1977). The activation of the Type II kinases also results in increased phosphorylation of regulatory subunit. The phosphorylated regulatory subunit reassociates more slowly with the catalytic subunit than the dephosphoform of the regulatory subunit which acts to prolong the effect of cyclic AMP (Rosen *et al.*, 1977).

Rosen *et al.* (1977) have recently reviewed the activation of the Type II protein kinase in cardiac muscle by cyclic AMP. Hosey and Tao (1977) have reviewed the role of protein kinases in membrane phosphorylation and Nimmo and Cohen (1977) the activation of protein kinases by hormones. These reviews should be consulted for more information on the role of cyclic AMP in mechanisms by which proteins are phosphorylated at available seryl or threonyl residues by cyclic AMP-dependent protein kinases.

1.7 COMPLEXITIES IN THE ROLE OF CYCLIC NUCLEOTIDES IN HORMONE ACTION: ACTIVATION OF GLYCOGEN PHOSPHORYLASE IN RAT LIVER

In fasting animals (including man) the main source of blood glucose is the liver (Cahill, 1971). The glucose released by the liver is derived from hepatic glycogenolysis until glycogen stores are exhausted and then glucose is formed from amino acids and lactate. The rate of glucose release by the liver appears to be regulated by both the level of blood glucose and the hormonal milieu (Cahill, 1971). Insulin appears to be the only physiologically important inhibitor of glucose release. In contrast, there are several hormonal stimulators of glucose release. Glucagon is currently favored as a prime regulator of glucose release. However, there is evidence in man that norepinephrine released from sympathetic nerve endings is important under certain condition (Brodows *et al.*, 1975).

The simple view that cyclic AMP is the regulator of hepatic glucose release whose formation is accelerated by glucagon or catecholamines and inhibited by insulin appears more untenable with the passage of time. There are marked species differences since in fish liver cells there appears to be only a cyclic AMP dependent activation of glycogenolysis and calcium is of little consequence (Birnbaum *et al.*, 1976). The original work on the role of cyclic AMP in hepatic glycogenolysis was done using dog liver in which both catecholamines and glucagon activate adenylate cyclase resulting in the formation of cyclic AMP which activates glycogen phosphorylase (Rall and Sutherland, 1958; Rall *et al.*, 1956, 1957). However, in man and the rat the activation of glycogen phosphorylase involves factors other than cyclic AMP.

Catecholamines appear to have two different types of effects which were originally classified by Ahlquist (1948) as α and β. Epinephrine has both α and β effects. In contrast, phenylephrine has primarily α effects and isoproterenol primarily β effects. The α effects of catecholamines are blocked by phentolamine, phenoxybenzamine and dihydroergotamine while a potent β-blocker is propranolol. The β effects of catecholamines are associated with the activation of adenylate cyclase and a rise in cyclic AMP. However, this does not mean that cyclic AMP is the sole mediator of β-catecholamine effects.

The α effects of catecholamines are not associated with a rise in cyclic AMP. In some cells α effects of catecholamines result in a drop in cyclic AMP and a rise in cyclic GMP. In these cells, the actions of cyclic AMP and GMP are antagonistic as are the α and β effects of catecholamines. It is important to realise that hepatocytes do not show competition between cyclic AMP and GMP. Epinephrine, which possesses both α and β effects, is the most potent activator of glycogenolysis or gluconeogenesis in rat liver cells.

The changes in hepatic cyclic GMP may reflect changes in free intracellular calcium. Indirect evidence for this was the finding that the divalent cation ionophore (A-23187) accelerated glycogenolysis and elevated cyclic GMP in rat liver cells (Pointer *et al.*, 1976). Glucagon and α-adrenergic agents also accelerate the uptake of labelled calcium by rat hepatocytes from the medium (Assimacopoulos-Jeannet *et al.*, 1977; Keppens *et al.*, 1977).

If α catecholamines increase cyclic GMP what does this mean with respect to regulation of hepatic glycogenolysis? There is substantial evidence that catecholamine-induced hepatic glycogenolysis in man (Day, 1975; Pilkington *et al.*, 1962) and the rat (Arnold and McAuliff, 1968; Gothelf and Ellis, 1974; Kennedy and Ellis, 1969) is predominantly an α response. It remained for Sherline *et al.*, (1972) to demonstrate using the perfused rat liver that phenylephrine increased glycogen phosphorylase without affecting cyclic AMP. Furthermore, phentolamine, an α-adrenergic antagonist, reduced the increase in phosphorylase activity but not cyclic AMP accumulation by epinephrine.

Sherline *et al.* (1972) suggested that the α effects of catecholamines might be secondary to vasoconstriction. However, in isolated rat hepatocytes where there are no α effects on vasoconstriction, the stimulation of hepatic gluconeogenesis by catecholamines was unrelated to cyclic AMP (Tolbert *et al.*, 1973). Subsequent studies from our laboratory (Birnbaum and Fain, 1977; Fain *et al.*, 1978; Pointer *et al.*, 1976) and elsewhere (Assimacopoulos-Jeannet *et al.*, 1977; Hutson *et al.*, 1976; Keppens *et al.*, 1977) have indicated that the α-adrenergic stimulation of glucose release and glycogen phosphorylase in isolated rat liver cells does not involve cyclic AMP.

If cyclic AMP is not the second messenger for the α stimulation of gluconeogenesis and glycogenolysis then what about cyclic GMP. We were initially excited by the finding that phenylephrine increased cyclic GMP in rat liver cells and this effect was blocked by phenoxybenzamine (Pointer *et al.*, 1976). According to the hypothesis of Berridge (1975) cyclic GMP might not be the second messenger but merely reflect a rise in cytosol calcium. In agreement with this hypothesis, we have never been able to obtain any affect of adding cyclic GMP or its derivatives on glycogenolysis by rat liver cells (Pointer and Fain, 1975).

The problem with involving cyclic GMP as a second messenger for α-catecholamine activation of glycogenolysis or gluconeogenesis is that a variety of other agents increase cyclic GMP. Carbachol increased cyclic GMP without affecting glycogenolysis while insulin increased cyclic GMP and inhibited hepatic glycogenolysis (Pointer *et al.*, 1976). The calcium ionophore A-23187 increased cyclic GMP and glycogenolysis in rat liver cells (Pointer *et al.*, 1976).

The rise in cyclic GMP due to α-catecholamines may reflect an elevation in calcium at the site of guanylate cyclase. However, if calcium is elevated by all the agents that increase cyclic GMP how can it be a regulator of glycogenolysis? Phosphorylase kinase in liver is a calcium-activated enzyme which could explain glucagon and catecholamine effects but not these of insulin and carbachol. We do not have any proof that the elevations in cyclic GMP due to insulin and carbachol are secondary to a rise in calcium. Furthermore, there are many pools of intracellular calcium and they are difficult to measure. About the only thing we can do at the moment is determine whether hormones affect total calcium content, calcium binding to components isolated from cells, influx and efflux of calcium or determine if extracellular calcium is required for the hormone effects.

Pointer *et al.* (1976) reported that in the absence of extracellular calcium (cells incubated in calcium-free buffer containing 1 mM EGTA) only the glycogenolytic effect of the cation ionophore A-23187 was abolished. The glycogenolytic effects of epinephrine and glucagon were variably reduced but never abolished in calcium-free buffer (Pointer *et al.*, 1976; Fain *et al.*, 1978).

Vasopressin activates glycogenolysis without affecting cyclic AMP accumulation. Kirk and Hems (1974) reported that in perfused rat livers vasopressin activated glycogenolysis without affecting cyclic AMP. Keppens and De Wulf (1975) demonstrated that glycogen phosphorylase was activated by vasopressin. Subsequently, Stubbs *et al.* (1976) found that the glycogenolytic action of vasopressin but not that of epinephrine or glucagon was abolished in calcium-free buffer. Keppens and De Wulf (1975) found that vasopressin and angiotensin activated phosphorylase but did not affect phosphorylase β-kinase, protein kinase or cyclic AMP.

Keppens *et al.* (1977) reported that, after incubation of liver cells for 45 min in calcium-free buffer, neither phenylephrine or vasopressin activated glycogen phosphorylase while the effect of glucagon was only reduced by 50%. Assimacopoulos-Jeannet *et al.* (1977) found that activation of phosphorylase by phenylephrine was reduced in the absence of extracellular calcium; However, we found as did Stubbs *et al.* (1976) that even after prolonged incubation in calcium-free buffer the effects of phenylephrine and epinephrine were not abolished (Fain *et al.*, 1900). The only effect of incubation in calcium-free buffer was to abolish the action of vasopressin. Probably the α-activation of glycogen phosphorylase does not involve cyclic nucleotides or require extracellular calcium. This does not rule out the possibility that α catecholamines might affect a redistribution of calcium within a discrete micro-environment within liver cells such as a plasma membrane pool.

Local changes in cyclic AMP might explain both α- and β-adrenergic activation since high concentrations of α-adrenergic antagonists block the increase in cyclic AMP due to phenylephrine seen in calcium-free buffer. However, the concentrations of α-catecholamines required to stimulate cyclic AMP accumulation are much higher than those required to activate glycogen phosphorylase. Furthermore the dose-response curve for α-activation of glycogenolysis by α-adrenergic agents was not potentiated by methyl xanthines (Birnbaum and Fain, 1977; Fain *et al.*, 1978).

Possibly low concentrations of glucagon and β-catecholamines act through unknown cyclic AMP-independent mechanisms to activate glycogen phosphorylase as shown in Fig. 1.1. Cyclic AMP could be an amplification signal which prolongs the action of glucagon and β-catecholamines on hepatic glycogen phosphorylase. Cherrington *et al.* (1977) have suggested that the only mechanism by which glucagon accelerates glycogenolysis is through cyclic AMP. They criticized the methodology used by Birnbaum and Fain (1977), but did not examine the response to low concentrations of glucagon in the absence of methyl xanthines. Cherrington *et al.* (1977) suggested that the dissociation between cyclic AMP and glycogen phosphorylase activation reported by Birnbaum and Fain (1977) was an artifact resulting from the experimental procedures. However, Okajima and Ui (1976) found a similar dissociation between cyclic AMP accumulation and glucagon action *in vivo*. Fain *et al.* (1978) have confirmed the findings of Birnbaum and Fain (1977). Furthermore, it has been known for some time that glucagon administration to dogs results in an increase in K^+ release which precedes any activation of glycogen phosphorylase (Frider *et al.*, (1964). However, it is still possible that all the effects of glucagon are secondary to an elevation of cyclic AMP in a compartment which is difficult to detect in the presence of a large amount of inactive cyclic AMP.

Ramachandran and Moyle (1977) have seen a similar dissociation between the effects of ACTH and steroidogenesis in the adrenal cortex and cyclic AMP. They found a derivative of ACTH which activated steroidogenesis under conditions in which there was no elevation of cyclic AMP. Rubin *et al.* (1972) had previously suggested that low concentrations of ACTH released 'trigger' Ca^{2+} bound to the cell membrane which was required in order to see any effects of the rise in cyclic AMP.

Mendelson *et al.* (1975) similarly reported that human chorionic gonadotropin stimulated steroidogenesis by Leydig cells at concentrations lower than those required to elevate cyclic AMP. They suggested that low concentrations of hormone worked by an unknown mechanism. More recently (Dufau *et al.*, (1977) examined the response to gonadotropin in the presence of 125 μM 1-methyl-3-isobutyl xanthine after 60 min incubation. There was a good correlation between hormone effects on cyclic AMP and steroidogenesis by purified Leydig cells in the presence of methyl xanthines. No studies were done in the absence of the methyl xanthine or after shorter time periods. Dufau *et al.* (1977) measured extracellular cyclic AMP and cyclic AMP bound to receptor protein as well as intracellular total cyclic AMP. They also measured free cyclic AMP binding sites and found no effect of hormone on binding affinity. Of the four assays for cyclic AMP, total intracellular cyclic AMP was the least sensitive but the other three procedures appeared to give the same results. In any case the measurement of cyclic AMP after 60 minutes incubation only in the presence of a high concentration of 1-methyl-3-isobutyl xanthine seems unusual. In order to obtain a definitive answer, the studies should be repeated in the presence of any methyl xanthine. The results do indicate that the measurement of cyclic AMP binding sites in cell extracts is more sensitive than the assay of total cyclic AMP.

In adipose tissue (Debons and Schwartz, 1961; Malbon *et al.*, 1978c; Vaughan, 1967) and heart (Hornbrook and Conrad, 1972; McNiel and Brody, 1968; Wildenthal, 1974) the stimulation of cyclic AMP accumulation by catecholamines is reduced by hypothyroidism. However, in rat liver cells we found that just the opposite occurs (Malbon *et al.*, 1978b). In liver cells from hypothyroid rats, the increase in cyclic AMP due to β-adrenergic agents was much greater than in liver cells from euthyroid rats. The sensitivity of liver glycogen phosphorylase activation to β-adrenergic agents was similarly enhanced by hypothyroidism. In contrast, the activation of both cyclic AMP accumulation and glycogen phosphorylase by glucagon was unaffected. The α-adrenergic stimulation of phosphorylase in rat liver cells was reduced in the hypothyroid state (Malbon *et al.*, 1978b).

Chan and Exton (1977) found similar effects of adrenalectomy. Again the results were unexpected because in most tissues the effects of catecholamines on metabolism are reduced in hypothyroid or adrenalectomized animals. The recent data can be reconciled if the physiological effects of catecholamines are mediated through α-adrenergic stimulation of glycogenolysis which was reduced by thyroidectomy (Malbon *et al.*, 1978b). These results support the hypothesis that the physiologically important regulation of hepatic glycogenolysis is mediated through α-adrenergic stimulation.

A mechanism for glucagon and catecholamine activation of hepatic glycogen phosphorylase is shown in Fig. 1.8. Low concentrations of glucagon may release 'trigger' calcium inside the liver cells which leads to the activation of phosphorylase and efflux of K^+.

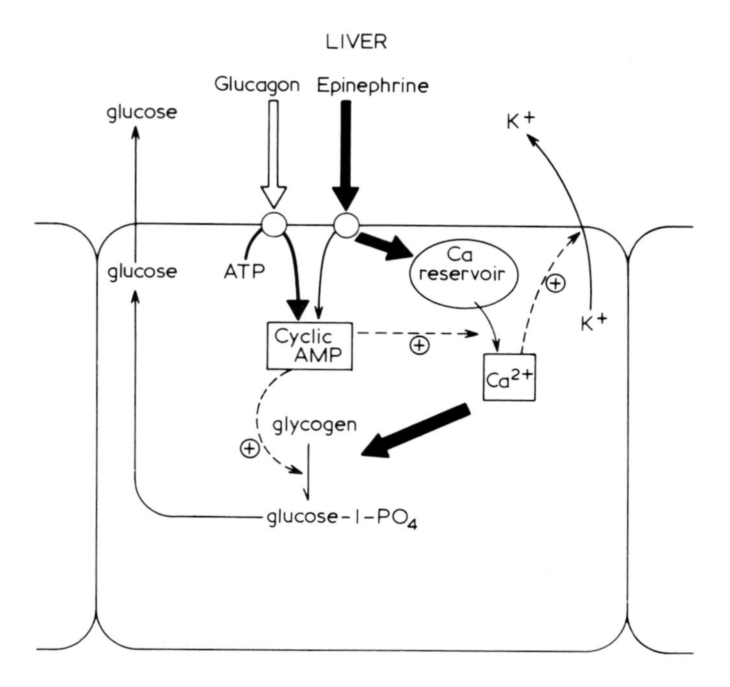

Fig. 1.8 Stimulation of glycogenolysis by epinephrine and glucagon in rat liver cells. The model suggests that both cyclic AMP and calcium are activators of glycogenolysis in liver cells. Epinephrine interacts with α-receptors and there is a breakdown of phosphatidylinositol resulting in the release of intracellular or 'trigger' calcium which activates glycogenolysis. In contrast, vasopressin (not shown) acts only by increasing the entry of calcium into liver cells. The epinephrine effect does not require extracellular calcium or any detectable rise in cyclic AMP. Glucagon is able to increase both cyclic AMP and release 'trigger' calcium. The glucagon-induced rise in cyclic AMP also releases intracellular stores of calcium. The model postulates that calcium and cyclic AMP are both activators of glycogen phosphorylase and that the primary effect of epinephrine is on the release of 'trigger' calcium. However, if there are no stores of intracellular calcium then epinephrine action is dependent on extracellular calcium whose entry can be accelerated by epinephrine.

Studies with dogs have shown that the *in vivo* administration of glucagon resulted in a release of K^+ which preceded the activation of glycogen phosphorylase (Finder *et al.*, 1964). Similar effects on K^+ release have also been observed in cats after epinephrine administration

Table 1.3 Comparison of the effects of various agents which activate rat liver glycogen phosphorylase

Agent	Cyclic AMP elevated	External Ca^{2+} required	Conclusion
A-23187 or vasopressin	No	Yes	Cyclic AMP-independent process which requires extracellular Ca^{2+}
Glucagon	Yes	No	Low concentrations elevate phosphorylase without requiring extracellular Ca^{2+} or a rise in cyclic AMP
β-Catecholamine	Yes	No	Low concentrations elevate phosphorylase without requiring extracellular Ca^{2+} or a rise in cyclic AMP
α-Catecholamine	Yes*	No	Unknown mechanism which may involve redistribution of intracellular Ca^{2+}

* Only at high concentrations in the presence of methyl xanthines

(Craig and Honig, 1963). Probably the difference between catecholamines and glucagon is that the former predominantly affects release of calcium while glucagon is able to activate both processes. The glycogenolytic effect of vasopressin differs from that of either glucagon or epinephrine in that it requires calcium uptake from the medium.

Recently, Kirk *et al.* (1977) found that vasopressin accelerated the uptake of ^{32}P into phosphatidyl inositol but not into other phospholipids. Within two minutes after the addition of vasopressin there was a ten-fold increase in ^{32}P incorporation. Epinephrine also accelerated ^{32}P incorporation into phosphatidylinositol but glucagon was inactive. The effect of vasopressin on ^{32}P incorporation into phosphatidyl inositol was seen in calcium-free buffer (Hems, personal communication) where there is no activation of phosphorylase by vasopressin (Kirk and Hems, 1974). These findings suggest that vasopressin increases the breakdown of plasma membrane-bound phosphatidyl inositol resulting in an increased entry of calcium into liver cells which mediates the activation of glycogen phosphorylase.

A summary of the three different mechanism by which rat liver glycogen phosphorylase can be activated is shown in Table 1.3. One mechanism involves entry of extracellular calcium which may be mediated directly by the addition of the divalent cation ionophore (A-23187). Vasopressin also appears to increase calcium entry into cells which may be secondary to increased breakdown of membrane phosphatidylinositol which opens up gates for calcium entry. The second mechanism is that for glucagon and β-catecholamines which can be mediated through cyclic AMP, but may also involve other factors. The third mechanism is the α-adrenergic activation of phosphorylase which does not involve cyclic AMP and is mediated through an unknown mechanism which may involve breakdown of phosphatidylinositol and release of 'trigger' Ca^{2+}.

1.8 ROLE OF PHOSPHATIDYLINOSITOL BREAKDOWN IN HORMONE ACTION

(a) *Thyrotropin and the thyroid gland*

Michell (1975; Michell *et al.*, 1976, 1977a,b) has postulated that many hormones increase phosphatidylinositol hydrolysis in the plasma membrane which results in an increase in membrane permeability to calcium. Thyrotropin increases phosphatidylinositol turnover and elevates cyclic AMP in the thyroid. However, the addition of dibutyryl cyclic AMP does not mimic the stimulation of TSH on phosphatidylinositol formation (Scott *et al.*, 1970). Furthermore, acetylcholine increases phosphatidylinositol synthesis but not cyclic AMP accumulation (Altman and Michell, 1974). Acetylcholine also elevates cyclic GMP (Yamashita and Field, 1972). The rise in cyclic GMP, the increased turnover of phosphatidylinositol and the rise in intracellular calcium appear to be linked.

Possibly, cyclic AMP is not the sole second messenger for thyrotropin. There may be a number of effects resulting from the interactions of thyrotropin with the plasma membrane. Batt and McKenzie (1976) found that thyrotropin hyperpolarized thyroid cells as did dibutyryl cyclic AMP and theophylline. Grollman *et al.* (1977) reported that the uptake of the lipophilic cation triphenylmethylphosphonium was increased by thyrotropin which was taken to reflect an alteration in the electrical potential of the membrane. The effect of thyrotropin on electrical potential preceded activation of adenylate cyclase and may reflect a primary action on the plasma membrane. It is not clear whether the changes in phospholipid metabolism due to thyrotropin have anything to do with the metabolic effects of this hormone. They can be mimicked by acetylcholine which does not affect thyroid hormone release.

(b) ***α-Adrenergic action on fat cells***

In fat cells, α-adrenergic agents (epinephrine in the presence of propranolol to block β effects) increase cyclic GMP (Fain and Butcher, 1976) and phosphatidylinositol turnover (Stein and Hales, 1972). However, there is no effect of α-adrenergic agents on fat cell lipolysis. In contrast, β-adrenergic agents activate lipolysis, increase cyclic AMP and the turnover of phosphatidylcholine (Stein and Hales, 1972). The rise in cyclic GMP due to catecholamines in rat fat cells can be mimicked by the calcium-ionophore (A-23187) which suggests that the α-stimulation of phosphatidylinositol turnover may be linked in some way to calcium (Fain and Butcher, 1976).

Lawrence and Larner (1977) found that there was an α-stimulation of glycogen phosphorylase and inhibition of glycogen synthase activity in fat cells. The α-adrenergic effect on glycogen metabolism in rat fat cells was dependent on extracellular calcium while that of β-adrenergic agents was not influenced by calcium (Lawrence and Larner, personal communication). These results support the hypothesis that there is an α-adrenergic activation of glycogen phosphorylase in rat fat cells, which is associated with a rise in cyclic GMP, phosphatidylinositol hydrolysis and a requirement for extracellular calcium.

One of the problems is the assays used for determination of phospholipid turnover in the presence of hormones. Most investigators in the past have not measured breakdown of the phosphatidylinositol but rather the uptake of labelled phosphate, inositol or other precursors. Much confusion has arisen from experiments using indirect assays.

Direct assays of phosphatidylinositol breakdown are required. While it is possible to prelabel phospholipids and look for decrease in radioactivity, there are two problems. The first is that of detecting small changes in a large quantity as pointed out by Michell (1975). The second problem is that the labeled phosphatidylinositol may not behave in the same manner as the unlabeled phospholipid.

It is easy to label cells by incubation with ^{32}P or tritiated inositol and then examine the efflux of label as an index of phosphatidylinositol breakdown after the addition of stimulators of cellular function. This procedure has been employed in studies using isolated islets of Langerhans obtained by collagenase digestion of rat pancreas. The addition of concentrations of glucose which stimulated secretion of insulin resulted in a transient efflux of labeled phosphate (Freinkel *et al.*, 1974). However, it has been difficult to demonstrate the source of this phosphate. Studies with labeled inositol would seem more promosing because there are fewer sources of inositol in cells for release to the medium (Clements and Rhoten, 1976). The addition of glucose resulted in a release of labeled inositol and inositol phosphates to the medium along with a loss of labeled phosphatidylinositol which was seen within 10 minutes (Clements and Rhoten, 1976). The problem in utilizing the efflux of inositol as an indicator of P_1 breakdown is that only 15% of the label in isolated islets was present in phosphatidylinositol.

(c) ***Serotonin and salivary secretion***

In isolated blowfly salivary glands the secretion of saliva is accelerated by serotonin. The addition of cyclic AMP to the medium stimulates salivary secretion but not the uptake and release of calcium into the saliva (Berridge, 1975). In the presence of serotonin the incorporation of labeled inorganic phosphate into phosphatidic acid was increased while that into all other phospholipids was described (Fain and Berridge 1978). There was a decrease in inorganic phosphate and labeled inositol incorporation into phosphatidylinositol in the presence of serotonin. Why serotonin inhibits the synthesis of phosphatidylinositol in the salivary gland and increases that of phosphatidic acid remains to be elucidated.

In salivary glands prelabeled with inositol or ^{32}P an increase in breakdown of labeled phosphatidylinositol was seen in the presence of serotonin (Fain and Berridge, 1978). Inositol release to the medium was a sensitive index of phosphatidylinositol hydrolysis in fly salivary glands. In these glands, of the inositol taken up which cannot be removed by brief washing, at least 94% of the label is in phosphatidylinositol (Fain and Berridge, 1978). The measurement of inositol release to the medium and saliva is more sensitive than measuring a small reduction in labeled phosphatidylinositol. The release of inositol occured as rapidly as the secretion of saliva was increased by concentrations of serotonin which accelerated salivary secretion. The addition of more hormone resulted in a progressive increase in inositol release while saliva secretion plateaued.

Michell *et al.* (1976) have pointed out that the dose–response curves for the activation of phosphatidylinositol breakdown by cholinergic agents are remarkably similar to those for receptor binding. This supports the hypothesis that the breakdown of phosphatidylinositol is involved in the primary action of the hormones receptor complex. Michell *et al.* (1976,

1977a,b) have suggested that muscarinic cholinergic responses are due to a release of membrane-bound calcium secondary to increased breakdown of phosphatidylinositol bound to the membrane. The increase in intracellular calcium by release of membrane-bound calcium is usually accompanied by an increase in calcium influx into the cells. Cytosol Ca^{2+} may be the second messenger which regulates cyclic GMP, secretion and contraction.

In blowfly salivary glands cyclic AMP, cytosol Ca^{2+}, and salivary secretion are increased by serotonin (Berridge, 1975). The addition of cyclic AMP mimics the stimulation of salivary secretion by serotonin but not the elevation in transepithelial calcium flux due to the hormone (Berridge, 1975). Just as in other systems the increased breakdown of phosphatidylinositol is not mimicked by cyclic AMP and is seen in calcium-free medium. Furthermore, the breakdown of phosphatidylinositol is not seen after the addition of calcium-ionophore just as Michell *et al.* (1977a, Jones and Michell, 1975) found in the parotid gland and other systems.

An interesting difference between blowfly salivary glands and other systems is that during hormonal stimulation there is an increased breakdown of phosphatidylinositol without any compensatory increase in phosphatidylinositol formation. When the hormone is removed there is an increase in phosphatidylinositol formation if inositol is present. The failure to see resynthesis while hormone is present may be due to an inhibition by elevated cytosol Ca^{2+} of phosphatidylinositol synthesis.

We found that in the absence of any added inositol, isolated blowfly salivary glands incubated in buffer have a very low rate of phosphatidylinositol formation and a high rate of phosphatidylglycerol synthesis (Fain and Berridge, 1978). If inositol is added to the medium there was a diversion of CDP diglyceride from phosphatidylglycerol synthesis to phosphatidylinositol formation. This suggests that there is a relatively constant turnover of CDP diglyceride which is used for phosphatidylinositol formation in the presence of inositol but, in its absence, is converted to phosphatidylglycerol.

Freinkel *et al.* (1975) reported similar findings in isolated rat pancreatic islets. If the glucose concentration in the medium was elevated to levels which stimulated insulin release there was an increased incorporation of ^{32}P into all phospholipids and phosphatidylinositol accounted for 75–80% of the ^{32}P incorporation. In the presence of 0.55 mM inositol the formation of phosphatidylinositol increased to 85% of total ^{32}P

incorporation into phospholipids and the appearance of label in phosphatidylglycerol and CDP diglyceride was virtually abolished. In the studies of Freinkel *et al.* (1975) the increase in phosphatidylinositol formation could have been secondary to an increased turnover of labeled ATP or uptake of inorganic ^{32}P into the islets.

Eichberg *et al.* (1973) also found an increased incorporation of ^{32}P into phosphatidylinositol and phosphatidylglycerol after the addition of catecholamines to rat glands in organ culture. This increase was unlikely to be a result of some secondary effect on the specific activity of precursor ATP since phosphatidylethanolamine and phosphatidylcholine formation were unaffected by catecholamines. In the unstimulated pineals the incorporation of ^{32}P into these two phospholipids was greater than that into phosphatidylinositol.

Local anesthetics such as tetracaine or high concentrations of propranolol mimicked the ability of catecholamine to stimulate ^{32}P incorporation into phosphatidic acid, phosphatidylinositol, phosphatidylglycerol and CDP diglyceride (Eichberg *et al.*, 1973; Eichberg and Hauser, 1974). However propranolol does not stimulate melatonin formation in the pineal gland. These data suggest that the stimulation of phosphatidylinositol formation by catecholamines in the pineal gland may not be linked with melatonin formation. The addition of cyclic AMP analogs to pineal glands stimulates melatonin formation, but there is no stimulation of phosphatidylinositol labeling.

Why are increases in membrane-bound phosphatidylinositol breakdown and free intracellular calcium seen in many cells along with an elevation in cyclic GMP? In the thyroid, fat cell and pineal glands the elevations of cyclic GMP and calcium have no known function and the addition of cyclic AMP can mimic hormone action. However, in other cells there are effects of hormones which are not mimicked by cyclic AMP and involve an elevation of cytosol calcium and cyclic GMP. In some cells (salivary gland for example), the same hormone has both effects while, in others, different hormones are involved for each effect.

What about the relationship between phosphatidylinositol breakdown and calcium? In most systems the effects of hormones on inositol metabolism are unaffected by the absence of extracellular calcium and the elevation of intracellular calcium by a variety of agents does not increase phosphatidylinositol breakdown (Michell *et al.*, 1976; 1977a,b). This supports the view that the breakdown of phosphatidylinositol is closely linked to the hormone–receptor interaction which results in a release of trigger calcium and increased calcium influx.

Lectins can increase phosphatidylinositol metabolism in lymphocytes but unlike the effects of hormone in other systems the presence of extracellular calcium is required and the effect of lectins is duplicated by a calcium ionophore (Hawthorne *et al.*, 1977). Allan and Michell (1974) found a soluble enzyme in lymphocytes which hydrolyzed phosphatidylinositol and was activated by 0.7 μM calcium. Apparently there is a second mechanism for increasing the breakdown of phosphatidylinositol which operates in lymphocytes. In other cells, this mechanism may also amplify the initial signal. Hormones interact with the membrane and increase phosphatidylinositol breakdown which results in the release of trigger calcium. The trigger calcium elevates cytosol calcium which leads to further hydrolysis of phosphatidylinositol.

The increased turnover of phosphatidylinositol in the presence of hormones was first noticed by Hokin and Hokin in 1955. It remained for Michell to point out the physiological significance of this finding. It is easy to obtain membrane preparations which respond to hormones with an activation of adenylate cyclase. However, convincing evidence that the addition of hormones increases breakdown of phosphatidylinositol by plasma membranes has not yet been obtained. Demonstration of this in an appropriate cell-free system would lend support to the hypothesis that this reaction is fully as important in explaining hormone action at the membrane level as the known activation of adenylate cyclase.

Michell (1975) postulated that cyclic inositol 1,2-phosphate, which is the compound formed when phosphatidylinositol is cleaved by intracellular phospholipases to diglyceride, might function as a second messenger. This is an attractive hypothesis, but no evidence has yet been found which proves or disproves the possible role of cyclic inositol 1,2-phosphate. In most systems, this compound is rapidly split to give inositol-1-phosphate and inositol. In blowfly salivary glands we find that only free inositol is released to the medium or saliva but there is a substantial buildup of inositol phosphates intracellularly after the addition of serotonin (Fain and Berridge, 1978). There appears to be more cyclic inositol 1,2-phosphate than inositol-1-phosphate

While a continued breakdown of phosphatidylinositol is required in salivary glands for the gating of calcium entry there is no proof that this is related to the formation of inositol 1:2 cyclic phosphate. It is equally likely that the diglyceride formed serves to gate calcium entry by reacting with proteins involved in calcium uptake. After removal of the hormone there is a rapid decrease in phosphatidylinositol breakdown and calcium flux through the salivary gland.

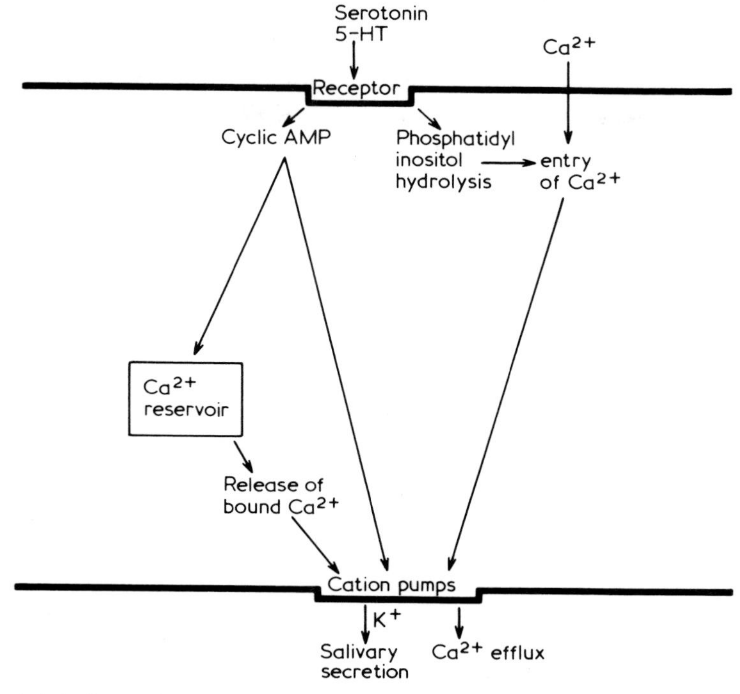

Fig. 1.9 Relationship between hormonal activation of cyclic AMP formation and phosphatidylinositol hydrolysis in the blowfly salivary gland in response to 5–HT (serotonin). The hormone–receptor complex activates both cyclic AMP formation and calcium flux. A single receptor is postulated which after reaction with serotonin activates both cyclic AMP formation by membrane-bound adenylate cyclase and phosphatidylinositol breakdown. Cyclic AMP activates the cation pumps responsible for fluid secretion. Elevation of cytosol Ca^{2+} is also required for fluid secretion. This Ca^{2+} can be derived for a time from the release of intracellular Ca^{2+} stores (Ca^{2+} reservoir) but continued secretion requires Ca^{2+} entry. Phosphatidylinositol breakdown in the membrane is linked in some unknown fashion to the gating of the Ca^{2+} entry.

The most exciting advance in linking calcium entry to phosphatidylinositol breakdown in the salivary gland was the finding that loss of responsiveness to serotonin with regard to calcium flux was related to a net decrease in phosphatidylinositol. After an hour exposure to 1 μM serotonin there was a net loss of over 50% of the labelled phosphatidylinositol and the ability of serotonin to increase calcium flux was markedly reduced. Incubation of the salivary glands depleted of phosphatidylinositol for one hour in the presence of 2 mM inositol (during a recovery period in the absence of hormone) resulted in a considerable increase in the rate of calcium flux due to addition of serotonin during a second incubation. In glands incubated for one hour without inositol there was no recovery of responsiveness with regard to calcium flux.

A model of the activation of salivary gland secretion and calcium flux is shown in Fig. 1.9. A unique feature of the fly salivary gland is that one hormone activates both processes. While there is no proof that the receptor for both effects is the same, this is the simplest hypothesis to explain the results. The hormone–receptor complex is postulated to activate adenylate cyclase which elevates cyclic AMP and phosphatidylinositol breakdown which increases entry of extracellular calcium. This is a good example of a system in which a hormone can activate membrane-bound adenylate cyclase and have other effects on membrane function which cannot be mimicked by cyclic AMP. Future work will undoubtedly result in more insights into the interrelationship between cyclic nucleotides, calcium and phosphatidylinositol breakdown.

REFERENCES

Ahlquist, R.P. (1948), *Am. J. Physiol.*, **153**, 586–600.

Allan, D. and Michell, R.H. (1974), *Biochem. J.*, **142**, 599–604.

Altman, M., Oka, H. and Field, J.B. (1966), *Biochim. biophys. Acta*, **116**, 586–588.

Angel, A., Desai, K.S. and Halperin, M.L., (1971), *J. Lipid Res.*, **12**, 203–211.

Appleman, M.M., Thompson, W.J. and Russell, T.R. (1973), *Adv. Cyclic Nucleotide Res.*, **3**, 65–98.

Armiento, M. d', Johnson, G. S. and Pastan, I. (1972), *Proc. natn. Acad. Sci. U.S.A.*, **69**, 459–462.

Arnold, A. and McAuliff, J.P. (1968), *Experientia*, **24**, 674.

Assimacopoulos-Jeannet, D., Blackmore, P.F. and Exton, J.H. (1977), *J. biol. Chem.*, **252**, 2662–2669.

Batt, R. and McKenzie, J.M. (1976), *Am. J. Physiol.*, **231**, 52–55.

Bennett, V., Mong, L. and Cuatrecasas, P. (1975), *J. Memb. Biol.*, **24**, 107–129.

Berridge, M.M. (1975), *Adv. Cyclic Nucleotide Res.*, **6**, 1–98.

Berridge, M. J. and Rapp, P.E. (1977), In: *Cyclic* 3′, 5′-*Nucleotides: Mechanisms of Action* (Cramer, H. and Schultz, J, eds), John Wiley, London, pp. 65–76.

Birnbaum, M. J. and Fain, J.N. (1977), *J. biol. Chem.*, **252**, 528–535.

Birnbaum, M.J., Schultz, J. and Fain, J.N. (1976), *Am. J. Physiol.*, **231**, 191–197.

Birnbaumer, L. and Rodbell, M. (1969), *J. biol. Chem.*, **244**, 3477–3482.

Bradshaw, R.A. and Frazier, W.A. (1977), *Curr. Topics Cell. Reg.*, **12**, 1–37.
Brivio-Haugland, R.P., Louis, S.L., Musch, K., Waldeck, N. and Williams, M.A. (1976), *Biochim. biophys. Acta*, **433**, 150–163.
Brodows, R.G., Pi-Sunyer, F.S. and Campbell, R.B. (1975), *Metabolism*, **24**, 617–624.
Brooker, G. (1973), *Science*, **182**, 933–934.
Brostrom, C.O., Huang, Y.C., Breckenridge, B. McL. and Wolff, D.J. (1975), *Proc. natn. Acad. Sci. U.S.A.*, **72**, 64–68.
Brostrom, M.A., Reimann, E.M., Walsh, D.A. and Krebs, E.G. (1970), *Adv. Enzyme Reg.*, **8**, 191–203.
Butcher, R.W. and Baird, C.E. (1968), *J. biol. Chem.*, **243**, 1713–1717.
Butcher, R.W. and Sutherland, E.W. (1962), *J. biol. Chem.*, **237**, 1244–1250.
Cahill, G. F., Jr. (1971), *Diabetes*, **20**, 785–799.
Cassell, D. and Selinger, Z. (1976), *Biochim. biophys. Acta*, **452**, 538–551.
Cassell, D. and Selinger, Z. (1977), *Proc. natn. Acad. Sci. U.S.A.*, **74**, 3307–3311.
Catt, K.J. and Dufau, M.L. (1977), *A. Rev. Physiol.*, **39**, 529–557.
Chan, T.F. and Exton, J.H. (1977), *Fedn. Proc. fedn. Am. Socs. exp. Biol.*, **36**, 384.
Chasin, M. and Harris, D.N. (1976), *Adv. Cyclic Nucleotide Res.*, **7**, 225–264.
Cherrington, A.D., Hundley, R.F., Dolgin, S. and Exton, J.H. (1977), *J. Cyclic Nucleotide Res.*, **3**, 263–273.
Cheung, W.Y. (1971), *J. biol. Chem.*, **246**, 2859–2869.
Chrisman, T.D., Garbers, D.L., Parks, M.A. and Hardman, J.G. (1975), *J. biol. Chem.*, **250**, 374–381.
Ciaraldi, T. and Marinetti, G.V. (1977), *Biochem. biophys. Res. Commun.*, **74**, 984–991.
Clark, R.B., Gross, R., Su, Y.F. and Perkins, J.P. (1974), *J. biol. Chem.*, **249**, 5296–5303.
Clements, R.S. and Rhoten, W.B. (1976), *J. clin. Invest.*, **57**, 684–691.
Craig, A. B., Jr. and Honig, C.R. (1963), *Am. J. Physiol.*, **205**, 1132–1138.
Cuatrecasas, P. (1974), *A. Rev. Biochem.*, **43**, 169–214.
Cuatrecasas, P. and Hollenberg, M.D. (1976), *Adv. Protein Chem.*, **30**, 251–451.
Cuatrecasas, P., Hollenberg, M.D., Chang, K. and Bennett, V. (1975), *Rec. Prog. Hormone Res.*, **31**, 37–94.
Czech, M.P. (1977), *A. Rev. Biochem.*, **46**, 359–384.
Dalton, C. and Hope, H.R. (1973), *Prostaglandins*, **4**, 641–651.
Dalton, C. and Hope, W.C. (1974), *Prostaglandins*, **6**, 227–242.
Davoren, P.R. and Sutherland, E.W. (1963), *J. biol. Chem.*, **238**, 3016–3023.
Day, J.L. (1975), *Metabolism*, **24**, 987–996.
Debons, A.F. and Schwartz, I.L. (1961), *J. Lipid Res.*, **2**, 86–91.
DeHaën, C. (1976), *J. Theor. Biol.*, **58**, 383–400.
DeRubertis, F.R. and Craven, P.A. (1976a), *J. biol. Chem.*, **251**, 4651–4658.
DeRubertis, F.R. and Craven, P.A. (1976b), *Science*, **193**, 897–899.

DeRubertis, F.R. and Craven, P.A. (1976c), *Metabolism,* **25**, 1113–1125.

Drummond, G. and Duncan, L. (1970), *J. biol. Chem.,* **245**, 976–983.

Dufau, M.L., Tsuruhara, T., Horner, K.A., Podesta, E. and Catt, K.J. (1977), *Proc. natn. Acad. Sci. U.S.A.,* **74**, 3419–3423.

Earp, H.S., Smith, P., Ong, S.H. and Steiner, A. (1977), *Proc. natn. Acad. Sci. U.S.A.,* **74**, 946–950.

Eichberg, J. and Hauser, G. (1974), *Biochem. biophys. Res. Commun.,* **60**, 1460–1467.

Eichberg, J., Shein, H.M., Schwartz, M. and Hauser, G. (1973), *J. biol. Chem.,* **248**, 3615–3622.

Engelhard, V.H., Esko, J.D., Storm, D.R. and Glaser, M. (1976), *Proc. natn. Acad. Sci. U.S.A.,* **73**, 4482–4486.

Fain, J.N. (1973a), *Pharmacol. Rev.,* **25**, 67–118.

Fain, J.N. (1973b), *Mol. Pharmacol.,* **9**, 595–604.

Fain, J.N. (1977), In: *Cyclic 3′, 5′-Nucleotides: Mechanisms of Action* (Cramer, H. and Schultz, J., eds), John Wiley, London, pp. 207–228.

Fain, J.N. (1978), In: *Biochemical Actions of Hormones,* (Litwack, G., ed.), Academic Press, New York, Vol. 7, in press.

Fain, J.N. and Berridge, M.J. (1978), *Biochem. Soc. Trans.,* in press.

Fain, J.N. and Butcher, F.R. (1976), *J. Cyclic Nucleotide Res.,* **2**, 71–78.

Fain, J.N., Dodd, A. and Novak, L. (1971), *Metabolism,* **20**, 109–118.

Fain, J.N., Li, S.H. and Butcher, F. R. (1978), unpublished.

Fain, J.N. and Loken, S.C. (1971), *Mol. Pharmacol.,* **7**, 455–464.

Fain, J.N., Pointer, R.H. and Ward, W.F., (1972), *J. biol. Chem.,* **247**, 6866–6872.

Fain, J.N., Psychoyos, S., Czernik, A.J., Frost, S. and Cash, W.D. (1973), *Endocrinology,* **93**, 632–639.

Fain, J.N. and Saperstein, R., (1970), *Hormone and Metabolic Research* Supplementum II Adipose Tissue Regulation and Function pp. 20–27, Georg Thieme, Stuttgart.

Fain, J.N. and Shepherd, R.E. (1975), *J. biol. Chem.,* **250**, 6586–6592.

Fain, J.N. and Shepherd, R.E. (1977), *J. biol. Chem.,* **252**, 8066–8070.

Fain, J.N. and Shepherd, R.E. (1978), In: *Hormones and Energy Metabolism,* (Klachko, D. M., Anderson, R.R., Heimberg, M., eds), Plenum Press, New York.

Fain, J.N., Shepherd, R.E., Malbon, C.C. and Moreno, F.J. (1978), In: *The Physiology of Lipid and Lipoprotein in Health and Disease,* American Physiological Society.

Fain, J.N. and Wieser, P.B. (1975), *J. biol. Chem.,* **250**, 1027–1034.

Ferreira, S.H. and Vane, J.R. (1974), *A. Rev. Pharmacol.,* **14**, 57–73.

Finder, A.G., Bayme, T. and Shoemaker, W.C. (1964), *Am. J. Physiol.,* **206**, 738–742.

Flores, J. and Sharp, G.W.G. (1975), *J. clin. Invest.,* **56**, 1345–1349.

Fredholm, B.B. and Hedqvist, P. (1975), *Biochem. Pharmacol.,* **24**, 61–66.

Freinkel, N., El Younsi, C. and Dawson, R.M.C. (1975), *Eur. J. Biochem.*, **59**, 245–252.

Freinkel, N., El Younsi, C., Bonnar, J. and Dawson, R.M.C. (1974), *J. clin. Invest.*, **54**, 1179–1189.

George, W.J., Polson, J.B., O'Toole, A.G. and Goldberg, N.D. (1970), *Proc. natn. Acad. Sci. U.S.A.*, **66**, 398–403.

Gerisch, G. and Wick, U. (1975), *Biochem. biophys. Res. Commun.*, **65**, 364–370.

Gill, D.M. and King, C.A. (1975), *J. biol. Chem.*, **250**, 6424–6432.

Gill, G.N. and Garren, L.D. (1970), *Biochem. biophys. Res. Commun.*, **39**, 335–343.

Gilman, A.G. (1970), *Proc. natn. Acad. Sci. U.S.A.*, **67**, 305–312.

Ginsberg, E., Solomon, D., Sreevalson, T. and Freese, E. (1973), *Proc. natn. Acad. Sci. U.S.A.*, **70**, 2457–2461.

Glossman, H. and Struck, C.J. (1977), *N.Z. Arch. Pharmacol.*, **299**, 175–185.

Goldberg, N.D. and Haddox, M.K. (1977), *A. Rev. Biochem.*, **46**, 823–896.

Goldberg, N.D., Haddox, M.K., Dunham, E., Lopez, C. and Hadden, J.W. (1974), In: *Control of Proliferation in Animal Cells*, (Clarkson, B. and Baserga, R., eds), Cold Spring Harbor Laboratory, pp. 609–625.

Gorman, R.R. (1975), *J. Cyclic Nucleotide Res.*, **1**, 1–9.

Gothelf, B. and Ellis, S. (1974), *Proc. Soc. exp. Biol. Med.*, **147**, 259–262.

Grollman, E.F., Lee, G., Ambesi-Impiombato, F. S., Meldolesi, M.F., Aloj, S.M., Coon, H.G., Kaback, H.R. and Kohn, L.D. (1977), *Proc. natn. Acad. Sci. U.S.A.*, **74**, 2352–2356.

Haddox, M.K., Furcht, L.T., Gentry, S.R., Moser, M.E., Stephenson, J.H. and Goldberg, N.D. (1976), *Nature*, **262**, 146–148.

Haga, T., Rosse, E.M., Anderson, H.J. and Gilman, A.G. (1977), *Proc. natn. Acad. Sci. U.S.A.*, **74**, 2016–2020.

Hamberg, M. and Samuelsson, B. (1974), *Proc. natn. Acad. Sci. U.S.A.*, **71**, 3400–3404.

Hawthorne, J.N., Pickard, M.R. and Griffin, H.D. (1977), *Biochem. Soc. Trans.*, **5**, 514–517.

Henneberry, R.C., Smith, C.C. and Tallman, J.F. (1977), *Nature*, **268**, 252–254.

Ho, R.J., Bomboy, J.D., Wasner, H.K. and Sutherland, E.W. (1975), *Meth. Enzym.*, **39**, 431–438.

Ho, R.J. and Sutherland, E.W. (1971), *J. biol. Chem.*, **246**, 6822–6827.

Hokin, M.R. and Hokin, L.E. (1955), *Biochim. biophys. Acta*, **18**, 102–110.

Holmgren, J., Lonnroth, I., Mansson, J.E. and Svennerholm, L. (1975), *Proc. natn. Acad. Sci. U.S.A.*, **72**, 2520–2524.

Hong, S.L. and Levine, L. (1976), *J. biol. Chem.*, **251**, 5814–5816.

Hornbrook, R.K. and Conrad, A. (1972), *Biochem. Pharmacol.*, **21**, 897–907.

Hosey, M.M. and Tao, M. (1977), *Curr. Topics Memb. Trans.*, **9**, 233–320.

Hutson, N.J., Brumley, T., Assimacopoulos, F.D., Harper, C. and Exton, J.H. (1976), *J. biol. Chem.*, **251**, 5200–5208.

Illiano, G.P. and Cuatrecasas, P. (1971), *Nature New Biol.*, **234**, 72–74.

Illiano, G.P., Tell, G.P.E., Siegel, M.I. and Cuatrecasas, P. (1973), *Proc. natn. Acad. Sci. U.S.A.*, **70**, 2443–2447.

Insel, P.A., Maguire, M.E., Gilman, A.G., Bourne, H.R., Coffino, P. and Melmon, K.L. (1976), *Mol. Pharmacol.*, **12**, 1062–1069.

Ismail, N.A., El Denshary, E.S.M. and Montague, W. (1977), *Biochem. J.*, **164**, 409–413.

Jones, L.M. and Michell, R.H. (1975), *Biochem. J.*, **148**, 479–485.

Kakiuchi, S. and Yamazaki, R. (1970), *Biochem. biophys. Res. Commun.*, **41**, 1104–1110.

Kebabian, J.W., Zatz, M., Romero, J.A. and Axelrod, J. (1975), *Proc. natn. Acad. Sci. U.S.A.*, **72**, 3735–3739.

Keppens, S. and De Wulf, H. (1975), *FEBS Letters*, **51**, 29–32.

Keppens, S., Van den Heede, J.R. and De Wulf, H. (1977), *Biochim. biophys. Acta*, **496**, 448–457.

Kennedy, B.L. and Ellis, S. (1969), *Arch. Int. Pharmacodyn. Therap.*, **177**, 390–406.

Kimura, H., Mittal, C. and Murad, F., (1975), *Nature*, **257**, 700–702.

Kimura, H. and Murad, F. (1975), *J. biol. Chem.*, **250**, 4810–4817.

Kimura, N. and Nagata, N. (1977), *J. biol. Chem.*, **252**, 3829–3835.

Kirk, C.J. and Hems, D. (1974), *FEBS Letters*, **47**, 128–131.

Kirk, C.J., Verrinder, T.R. and Hems, D.A. (1977), *FEBS Letters*, **83**, 267–271.

Kumon, A., Yamamura, H. and Nichizuka, Y. (1972), *Biochem. biophys. Res. Commun.*, **41**, 1290–1297.

Kuo, J.F. and Greengard, P. (1969), *Proc. natn. Acad. Sci. U.S.A.*, **64**, 1349–1355.

Kurosky, A., Markel, D.E. and Peterson, J.W. (1977), *J. biol. Chem.*, **252**, 7257–7264.

Lai, C.Y. (1977), *J. biol. Chem.*, **252**, 7249–7256.

Langan, T.A. (1973), *Adv. Cyclic Nucleotide Res.*, **3**, 99–153.

Lawrence, J.C. and Larner, J. (1977), *Mol. Pharmacol.*, **13**, 1060–1075.

Levey, G.S. (1971), *J. biol. Chem.*, **246**, 7405–7410.

Levey, G.S. (1973), *Recent Prog. Hormone Res.*, **29**, 361–386.

Levinson, S.L. and Blume, A.J. (1977), *J. biol. Chem.*, **252**, 3766–3774.

Limbird, L.E. and Lefkowitz, R.J. (1977), *J. biol. Chem.*, **252**, 799–802.

Londos, C. and Preston, M.S. (1977), *J. biol. Chem.*, **252**, 5951–5956.

Londos, C., Salomon, Y., Lin, M.C., Harwood, J.P., Schramm, M., Wolff, J. and Rodbell, M. (1974), *Proc. natn. Acad. Sci. U.S.A.*, **71**, 3087–3090.

Loten, E.G. and Sneyd, J.G.T. (1970), *Biochem. J.*, **120**, 187–193.

Malbon, C.C. and Fain, J.N. (1978), unpublished.

Malbon, C.C., Hert, R.C. and Fain, J.N. (1978a), *J. biol. Chem.*, **253**.

Malbon, C.C., Li, S. and Fain, J.N. (1978b), unpublished.

Malbon, C.C., Moreno, F.J., Cabelli, R.J. and Fain, J.N. (1978c), *J. biol. Chem.*, **253**, 671–678.

Malgieri, J., Shepherd, R.E. and Fain, J.N. (1975), *J. biol. Chem.*, **250**, 6593–6598.

Manganiello, V., Murad, F. and Vaughan, M. (1971), *J. biol. Chem.*, **246**, 2195–2202.

McNiel, J.H. and Brody, T.M. (1968), *J. Pharmacol. exp. Therap.*, **161**, 40–46.
Mendelson, C., Dufau, M. and Catt, K.J. (1975), *J. biol. Chem.*, **250**, 8818–8823.
Michell, R.H. (1975), *Biochim. biophys. Acta*, **415**, 81–147.
Michell, R.H., Jafferji, S.S. and Jones, L.M. (1976), *FEBS Letters*, **69**, 1–5.
Michell, R.H., Jafferji, S.S. and Jones, L.M. (1977a), *Adv. exp. Biol., Med.*, **83**, 447–465.
Michell, R.H., Jones, L.M. and Jafferji, S.S. (1977b), *Biochem. Soc. Trans.*, **5**, 77–81.
Moss, J., Manganiello, V. C. and Vaughan, M. (1976), *Proc. natn. Acad. Sci. U.S.A.*, **73**, 4424–4427.
Mukherjee, C., Caron, M.G. and Lefkowitz, R.J. (1975), *Proc. natn. Acad. Sci. U.S.A.*, **72**, 1945–1949.
Mukherjee, C. and Lefkowitz, R.J. (1976), *Proc. natn. Acad. Sci. U.S.A.*, **73**, 1494–1498.
Neer, E.J. (1973), *J. biol. Chem.*, **248**, 3742–3744.
Nimmo, H.G. and Cohen, P. (1977), *Adv. Cyclic Nucleotide Res.*, **8**, 145–266.
Okajima, F. and Ui, M. (1976), *Arch. biochem. Biophys.*, **175**, 549–557.
Olsson, R.A., Davis, C.J., Khouri, E.M. and Patterson, R.E. (1976), *Circulation Res.*, **39**, 93–98.
Orly, J. and Schramm, M. (1976), *Proc. natn. Acad. Sci. U.S.A.*, **73**, 4410–4414.
Oye, I. and Sutherland, E.W. (1966), *Biochim. biophys. Acta*, **127**, 347–354.
Pastan, I., Roth, J. and Macchia, V. (1966), *Proc. natn. Acad. Sci. U.S.A.*, **56**, 1802–1809.
Pawlson, L.G., Lovell-Smith, C.J., Manganiello, V.C. and Vaughan, M. (1974), *Proc. natn. Acad. Sci. U.S.A.*, **71**, 1639–1642.
Perkins, J.P. (1973), *Adv. Cyclic Nucleotide Res.*, **3**, 1–64.
Perkins, J.P., Moore, M.M., Kalisher, A. and Su, Y.F. (1975), *Adv. Cyclic Nucleotide Res.*, **5**, 641–660.
Pfeuffer, T. (1977), *J. biol. Chem.*, **252**, 7224–7234.
Pfeuffer, T. and Helmreich, E.J.M. (1975), *J. biol. Chem.*, **250**, 867–876.
Pilkington, T.R.E., Lowe, R.D., Robinson, B.F. and Titterington, E. (1962), *Lancet*, **2**, 316–317.
Pointer, R.H., Butcher, F.R. and Fain, J.N. (1976), *J. biol. Chem.*, **251**, 2987–2992.
Pointer, R.H. and Fain, J.N. (1975), unpublished studies.
Rajerison, R., Marchetti, J., Roy, C., Boeckaert, J. and Jard, S. (1974), *J. biol. Chem.*, **249**, 6390–6400.
Rall, T.W. and Sutherland, E.W. (1958), *J. biol. Chem.*, **232**, 1065–1076.
Rall, T.W., Sutherland, E.W. and Berthet, J. (1957), *J. biol. Chem.*, **224**, 463–475.
Rall, T.W., Sutherland, E.W. and Wosilait, W.D. (1956), *J. biol. Chem.*, **218**, 483–495.
Ramachandran, J. and Moyle, W.R. (1977), In: *Endocrinology* (James, V.H.T., ed.), Vol. 1, Excerpta Medica, Amsterdam, pp. 520–525.
Rapp, P.E. and Berridge, M.J. (1977), *J. Theor. Biol.*, **66**, 497–525.
Rasmussen, H. (1970), *Science*, **170**, 404–412.
Rasmussen, H. and Goodman, D.B.P. (1977), *Physiol. Rev.*, **57**, 421–509.
Rendell, M.S., Rodbell, M. and Berman, M. (1977), *J. biol. Chem.*, **252**, 7909–7912.

Robison, G.A., Butcher, R.W. and Sutherland, E.W. (1971), *Cyclic AMP,* Academic Press, New York.

Rodbell, M. (1965), *Ann. N.Y. Acad. Sci.,* **131**, 302–314.

Rodbell, M., Birnbaumer, L. and Pohl, S.L. (1970), *J. biol. Chem.,* **245**, 718–722.

Rodbell, M., Birnbaumer, L., Pohl, S.L. and Krans, H. M. J. (1971), *J. biol. Chem.,* **246**, 1877–1882.

Rosen, O.M., Rangel-Aldao, R. and Ehrlichman, J. (1977), *Curr. Topics Cell. Reg.,* **12**, 39–74.

Ross, E.M. and Gilman, A.G. (1977), *J. biol. Chem.,* **252**, 6966–6969.

Rubin, R.P., Carchman, R.A. and Jaanus, S.D. (1972), *Nature, New Biol.,* **240**, 150–152.

Sattin, A. and Rall, T.W. (1970), *Molec. Pharmacol.,* **6**, 13–23.

Schimmer, B.P., Ueda, K. and Sato, G.H. (1968), *Biochem. biophys. Res. Commun.,* **32**, 806–810.

Schramm, M., Orly, J., Eimerl, S. and Korner, M. (1977), *Nature,* **268**, 310–313.

Schwabe, U., Berndt, S. and Ebert, R. (1972), *Arch. Pharmacol.,* **273**, 62–74.

Schwabe, U. and Ebert, R. (1972), *Arch. Pharmacol.,* **274**, 287–298.

Schwabe, U. and Ebert, R. (1974), *Arch. Pharmacol.,* **282**, 33–44.

Schwabe, U., Ebert, R. and Erbler, H.C. (1973), *Arch. Pharmacol.,* **276**, 133–148.

Schultz, G., Hardman, J.G., Schultz, K., Baird, C.E. and Sutherland, E.W. (1973), *Proc. natn. Acad. Sci. U.S.A.,* **70**, 3889–3893.

Scott, T.N., Freinkel, N., Klein, J.H. and Nitzan, M. (1970), *Endocrinology,* **87**, 854–863.

Scow, R.O. (1965), *Handbook of Physiology,* Adipose Tissue Section 5, American Physiological Society, Bethesda, pp. 437–453.

Sherline, P., Lynch, A. and Glinsman, W.H. (1972), *Endocrinology,* **91**, 680–690.

Shier, W.T., Baldwin, J.H., Nilsen-Hamilton, M., Hamilton, R.T. and Thonassi, N.M. (1976), *Proc. natn. Acad. Sci. U.S.A.,* **73**, 1586–1590.

Siegel, M.I., Puca, G.A. and Cuatrecasas, P., (1976), *Biochim. biophys. Acta,* **438**, 310–323.

Singer, S.J. and Nicolson, G.L. (1972), *Science,* **175**, 720–731.

Skolnick, P. and Daly, J.W. (1977), In: *Cyclic 3′, 5′-Nucleotides: Mechanisms of Action,* (Cramer, H. and Schultz, J., eds), pp. 289–315.

Sonenberg, M. and Schneider, A.S. (1977), In: *Receptors and Recognition* Series A, Vol. 4, pp. 1–73, Chapman and Hall, London.

Stein, J.M. and Hales, C.N. (1972), *Biochem. J.,* **128**, 531–541.

Steiner, A.L., Parker, C.W. and Kipnis, D.M. (1972), *J. biol. Chem.,* **247**, 1106–1113.

Streeto, J.M. (1969), *Metabolism,* **18**, 969–973.

Stubbs, M., Kirk, C.J. and Hems, D.A. (1976), *FEBS Letters,* **69**, 199–202.

Sutherland, E.W. and Rall, T.W. (1958), *J. biol. Chem.,* **232**, 1077–1091.

Tao, M., Salas, M.L. and Lipmann, F. (1970), *Proc. natn. Acad. Sci. U.S.A.,* **67**, 408–414.

Tolbert, M.E.M., Butcher, F.R. and Fain, J.N. (1973), *J. biol. Chem.,* **248**, 5866–5692.

Trost, T. and Stock, K. (1977), *Arch. Pharmacol.*, **299**, 33–40.
Van Heyningen, S. and King, C.A. (1975), *Biochem. J.*, **146**, 269–271.
Van Heyningen, S. (1977), *Biol. Rev.*, **52**, 509–549.
Van Heyningen, W.E. (1973), *Arch. Pharmacol.*, **276**, 289–295.
Van Heyningen, W.E., Carpenter, C.C. Jr., Pierce, N.F. and Greenough, W.B. III (1971), *J. Infect. Dis.*, **124**, 415–418.
Van Sande, J., Decoster, C. and Dumont, J.E. (1975), *Biochem. biophys. Res. Commun.*, **62**, 168–175.
Vaughan, M. (1967), *J. Clin. Invest.*, **46**, 1482–1491.
Wang, J.H. (1977), In: *Cyclic* 3′, 5′*-Nucleotides: Mechanisms of Action* (Cramer, H. and Schultz, J., eds), John Wiley, London, pp. 37–56.
Weiss, B. and Hart, W.N. (1977), *A. Rev. Pharmacol.*, **17**, 441–477.
Wells, J.N. and Hardmann, J.G. (1977), *Adv. Cyclic Nucleotide Res.*, **8**, 119–144.
Wildenthal, K. (1974), *J. Pharmacol. exp. Therap.*, **190**, 272–279.
Williams, L.T., Lefkowitz, R.J., Watanabe, A.M., Hathaway, D.R. and Besch, H.R. Jr. (1977), *J. biol. Chem.*, **252**, 2787–2789.
Yamamura, H., Lad, P.M. and Rodbell, M. (1977), *J. biol. Chem.*, **252**, 7964–7966.
Yamashita, K. and Field, J.B. (1972), *J. biol. Chem.*, **247**, 7062–7066.

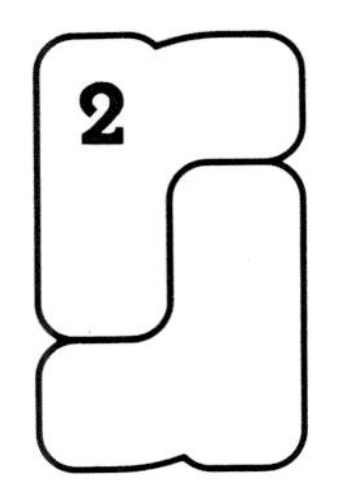

Reconstitution of Biological Membranes

GERA D. EYTAN
Department of Biology,
Technion- Israel Institute of Technology, Haifa
and
BARUCH I. KANNER
Department of Medical Biochemistry, Hadassa Medical School,
The Hebrew University, Jerusalem

Receptors and Recognition, Series A, Volume 6
Edited by P. Cuatrecasas and M.F. Greaves
Published in 1978 by Chapman and Hall, 11 New Fetter Lane, London EC4P 4EE

INTRODUCTION

Biological membranes are currently envisaged as a fluid lipid bilayer with the membrane proteins freely floating in it (Singer and Nicolson, 1972). Most of these proteins are involved in membrane-linked functions such as solute translocation. Detailed characterisation of membrane processes requires solubilisation and purification of the proteins. However, upon disruption of the membrane structure, processes such as solute transport can no longer be observed. Recently, techniques have become available to reconstitute membrane functions, by incorporating the proteins into lipid model systems (Racker, 1972a; Racker *et al.*, 1975a). This approach represents a new area in membrane biochemistry. In the present review we will illustrate the principles underlying membrane reconstitution using a few selected examples. We shall describe in detail some aspects of reconstitution of microsomal electron transport and of proton translocation catalyzed by cytochrome oxidase. Special emphasis will be put on solubilisation, purification and reconstitution of inner mitochondrial membrane proteins as most of this membrane's proteins have been successfully reconstituted. This has allowed studies of the interaction of the various proteins in isolated well-defined model systems. For extensive reviews covering the details of the multitude of systems reconstituted, we refer the reader to a few excellent reviews (Montal, 1976; Korenbrot, 1977; Racker *et al.*, 1975a).

2.1 EXPERIMENTAL LIPID MODELS

2.1.1 Monolayers

Phospholipids are amphiphilic molecules, their headgroups are hydrophilic and their fatty acyl groups hydrophobic. At surfaces of aqueous solutions they form a monolayer of molecules with their head groups in the water and their apolar moieties exposed to the air. Monolayers proved very useful in studies of molecular organization of phospholipids and other lipids as a function of parameters such as temperature, pressure, nature of the phospholipids and electrostatic fields (Montal, 1976; Demel, 1974;

Phillips, 1972). Monolayers, in contrast to biological membranes, have aqueous phase next to one face only and thus their relevance to research of transport mechanisms is limited. Nevertheless they served in affinity studies of membrane proteins to phospholipid classes (Boguslavsky *et al.*, 1975; Colacicco, 1970; Klappauf and Schubert, 1977; Rothfield and Fried, 1976; van Zoelen, *et al.*, 1977). London *et al.* (1973, 1974) injected ^{131}I-labelled brain membrane proteins into the buffer beneath the monolayers of different compositions and assessed the affinity of the proteins to the lipids by monitoring the concentration of radioactivity next to the surface. The nature of the protein–lipid interactions were studied by measuring surface pressure and susceptibility of the proteins to proteases. The A_1 basic protein was bound specifically to cerebroside sulphate while the 'Folch-Lees Protein' was bound preferentially to cholesterol.

2.1.2 Liposomes

Upon hydration of phospholipids, they assume the bilayer structure. Mechanical shaking induces them to form a concentric mutilayer particle reminiscent of an onion-like structure (Hauser *et al.*, 1972; Gregoriades *et al.*, 1971). For studies of ion-translocating membrane proteins the multilayered vesicles have two major disadvantages: (a) their surface area is relatively low, as most of their phospholipid bilayers are buried within them and (b) evaluation of transport kinetics is made almost impossible by the multitude of compartments bounded by the consecutive membranes.

The multilayered vesicles can be transformed into small single-layered vesicles called liposomes by sonic disruption (Bangham, 1968; Bangham *et al.*, 1974; Tyrrell *et al.*, 1976; Papahadjopoulos and Kimelberg, 1973; Barenholz *et al.*, 1977). Alternatively, unilamellared liposomes were formed by dilution of phospholipids dissolved in ethanol (Batzri and Korn, 1973) or by dialysis of phospholipid solution in detergent, i.e. cholate (Racker, 1972a, Brunner *et al.*, 1976). Residual contaminating multilayered vesicles can be removed by centrifugation or Sepharose chromatography leaving behind a strikingly homogeneous population of liposomes (Huang, 1969). These liposomes have been the subject of extensive research for the last decade and several excellent reviews have been published (Bangham, 1968; Bangham *et al.*, 1974; Tyrrell *et al.*, 1976; Papahadjopoulos and Kimelberg, 1973). Therefore, we shall mention only a few characteristics of these liposomes relevant to membrane reconstitution: the best-studied liposomes are those prepared from egg phosphatidylcholine (Huang, 1969; Sheetz and Chan, 1972; Aune *et al.*, 1977). They have a molecular weight of approximately 2×10^6, corresponding to 2600 phospholipid molecules per liposome. The outer diameter of these liposomes is 210 Å and the inner diameter is 170 Å. The outer surface area of these liposomes is 2.6 times the inner surface area. In small liposomes containing two or more phospholipids, the high degree of curvature induces compositional asymmetry and different molecular packing arrangements between the inner and outer faces of the particles (Spiker and Levin, 1976; Berden *et al.*, 1975; Michaelson *et al.*, 1974; Huang *et al.*, 1974; Brunner *et al.*, 1976). For example, in liposomes containing both phosphatidylethanolamine and phosphatidylcholine the latter is preferentially concentrated in the outer monolayer. In liposomes containing phosphatidylglycerol and phosphatidylcholine, the latter is preferentially inside. In addition, liposomes can undergo lateral phase separations resulting in co-existence of domains rich in different species of lipids (Shimshick and McConnell, 1973; Ohnishi and Ito, 1973; Papahadjopoulos *et al.*, 1974). Phase separations may be induced by selectively aggregating and solidifying acidic phospholipids with divalent cations or polylysine (Hartmann *et al.*, 1977; Papahadjopoulos *et al.*, 1974; Galla and Sackmann, 1975). Liposomes undergo thermal phase transitions from an orderly solid structure existing below the transition temperature to a fluid smectic arrangement existing at higher temperatures. Lateral phase separation occurs as well at the transition temperature, when solid domains co-exist in the liposomes with liquid regions. Liposomes are relatively impermeable to ions and to hydrophilic molecules (Demel *et al.*, 1968, 1972; Bangham *et al.*, 1974).

Small single-layered liposomes proved very useful in reconstitution of membrane proteins. Several methods for the functional introduction of membrane proteins* into single-walled lipid vesicles are now available.

(a) ***Cholate dialysis***

The protein together with phospholipids are dissolved using detergent, mostly cholate. The detergent is slowly removed by dialysis overnight (Kagawa and Racker, 1971; Hinkle *et al.*, 1972; Racker, 1972a). This method proved successful for all systems tested and thus is the method of choice for the exploration of new systems.

* The proteins are usually solubilised using detergents. The use of these detergents has been reviewed (Hellenius and Simons, 1975; Tanford and Reynolds, 1976).

(b) ***Cholate dilution***
The protein and phospholipids are dissolved as above with cholate. The mixture is incubated for 10 min at 0° C and aliquots are withdrawn and assayed. Reconstitution of functional proteoliposomes occurs upon dilution of the detergent into the reaction mixture (Racker *et al.*, 1975a; Serrano *et al.*, 1976). Reconstitution by cholate dilution has been found rapid, reproducible and less harmful than cholate dialysis, since exposure of the protein to the detergent is brief. Multiple samples can be assayed simultaneously.

(c) ***Sonication***
Proteoliposomes are formed by sonic irradiation of phospholipid together with the proteins preferably using a bath type sonicator (Racker, 1973). The method proved the best for some systems, such as bacteriorhodopsin, but prolonged sonication may be harmful in other cases (Barnerjee *et al.*, 1977).

(d) ***Incorporation into preformed liposomes***
Liposomes containing either low concentrations of detergents (such as lysolecithin or cholate) or acidic phospholipid are incubated with the isolated protein. Incorporation occurs within a few minutes (Eytan *et al.*, 1976a,b; Eytan and Racker, 1977).

A major drawback of single-walled proteoliposomes is their small inner volume, limiting their capacity. Recently, procedures for producing large single-layered liposomes have been described (Deamer *et al.*, 1976; Kremer *et al.*, 1977; Papahadjopoulos *et al.*, 1975; Reeves and Dowben, 1970). However, no protein has been incorporated into these particles. A feasible procedure for obtaining large proteoliposomes is fusion of regular small ones. The fusion of proteoliposomes is catalyzed by Ca^{2+} ions, presence of acidic phospholipids in the liposomes and osmotic pressure exerted across the liposomes' membranes, (Papahadjopoulos *et al.*, 1974; Miller and Racker, 1976; Miller *et al.*, 1976). The large proteoliposomes, up to 2 μm radius, should prove interesting for future investigation, as they are more analogous to biological membranes, being single-layered and with low curvature.

2.1.3 Planar lipid membranes

Transport of charged molecules across artificial membranes is best measured as electrical conductance. Mueller and Rudin (Mueller *et al.*, 1962) have developed for this purpose the planar lipid membrane system. They applied a drop of lipid solution in hydrocarbon to an aperture in a teflon partition separating two aqueous phases. The hydrocarbon and excess lipids were drawn to the hydrophobic teflon leaving across the aperture a lipid bilayer containing only residual entrapped solvent lenses. Montal (1974a,b) has suggested an alternative model system; a planar lipid membrane is formed by adjoining the hydrocarbon chains of two monolayers, originally at the surface of the two aqueous phases separated by the partition. The resultant bilayer lipid structure does not contain organic solvents. Planar lipid membranes served a pivot role in elucidating the mechanism of ionophore conductance as described recently in selected reviews (Fettiplace *et al.*, 1975; MacLaughlin and Eisenberg, 1975; Pressman, 1976). In the present review we will describe only briefly some recent approaches to membrane proteins function using planar lipid membrane (see also Montal, 1976).

2.2 RECONSTITUTION OF CYTOCHROME b_5 AND CYTOCHROME b_5 REDUCTASE

The hemoprotein cytochrome b_5 and the flavoprotein NADH – cytochrome b_5 reductase are members of the liver microsomal electron transfer chain which catalyzes NADH oxidation through the sequence 'NADH → cytochrome b_5 reductase → cytochrome b_5 → stearylCoA desaturase (Cohen and Estabrook, 1971; Oshino and Sato, 1971; Oshino *et al.*, 1971; Shimataka *et al.*, 1972; Strittmatter *et al.*, 1972). While the primary function of cytochrome b_5 is to transfer reducing equivalents from NADH cytochrome b_5 reductase to stearylCoA desaturase, it has been suggested that it may participate in NADH-linked cytochrome P_{450} catalyzed reactions as well (Cohen and Estabrook, 1971; Oshino *et al.*, 1971; Ozols *et al.*, 1976). Both NADH cytochrome b_5 reductase and cytochrome b_5 are amphipatic proteins containing a hydrophilic segment to which the prosthetic group is bound and a segment which is rich in hydrophobic amino acids and serves to bind the proteins to the microsomes. Cytochrome b_5 may be isolated in two forms: a protein with a molecular weight of 16 700 containing both the hydrophobic and hydrophilic segments and a polypeptide with a molecular weight of 10 700, containing only the hydrophilic segment. The first form is obtained with detergents, while the second form is a product of tryptic digestion (Spatz and Strittmatter, 1971). The complete amino acid sequence of the protein has been obtained (Ozols, 1970; Ozols and Gerard, 1977). Investigation of the isolated enzyme with X-ray diffraction and analytical centrifugation has indicated that both the hydrophobic and hydrophilic segments are folded into globular regions which are joined by a 30–40 Å long flexible region (Matthews *et al.*, 1971; Robinson and Tanford, 1975; Visser *et al.*, 1975).

Isolated intact cytochrome b_5 and NADH–cytochrome b_5 reductase may be bound to liver microsomes. Under saturating conditions, the maximal amount of extra cytochrome b_5 bound is equal to 20% of the weight and represents a 10-fold increase in the cytochrome b_5 content of the microsomes (Rogers and Strittmatter, 1974; Strittmatter *et al.*, 1972a,b). The binding occurs via the hydrophobic segment as the hydrophylic heme peptide does not bind to microsomes. The extra cytochrome b_5 bound is catalytically indistinguishable from the endogenous cytochrome b_5. Similarly, the maximal amount of NADH-cytochrome b_5 reductase that may be bound represents a 100-fold increase over the endogenous reductase amount and is functionally active. All the enzymes bound, as well as the endogenous proteins, are oriented on the outer surface of the microsomes. These observations led Strittmatter and his colleagues to suggest that cytochrome b_5 and the reductase do not form a stable complex in the membrane. They are randomly distributed on the external surface and undergo lateral diffusion within the membrane. Chemical interactions between the proteins occurs upon collisions of the catalytic segments of the enzymes. This model was further confirmed when the proteins were bound to liposomes.

Cytochrome b_5 reductase and the intact cytochrome b_5, but not the heme b_5 polypeptide, may be bound to phosphatidylcholine liposomes (Rogers and Strittmatter, 1975). Reconstitution of cytochrome b_5 vesicles is achieved by incubation of the enzyme with the liposomes. The excess cytochrome b_5 is removed either by chromatography or centrifugation into sucrose gradients. No detergents are added to the incorporation medium and the enzyme contains less than one detergent molecule per cytochrome b_5 molecule (Holloway and Katz, 1975). The incorporation occurs even into synthetic phosphatidylcholine liposomes at temperatures below the phospholipid melting point (Enoch *et al.*, 1977). The binding of the cytochrome is hydrophobic and does not involve electrostatic attraction as the liposomes used are neutral. Tanford and his colleagues (Robinson and Tanford, 1975; Visser *et al.*, 1975), who studied the binding characteristics of isolated cytochrome b_5 observed that the hydrophobic region of the enzyme is the part participating in binding both phospholipids and non-denaturing detergents. They suggested that the hydrophobic globule of the enzyme has a diffuse hydrophobic surface. The maximal cytochrome b_5 that may be bound to liposomes is 244 molecules per liposome, which represents a 50% excess of protein over lipid. Under these conditions approximately 82% of the liposome surface is covered by the cytochrome (Rogers and Strittmatter, 1975). Taking into account possible repulsion forces between the molecules, one can safely conclude that the limiting factor in binding of cytochrome b_5 to liposomes is the available surface area.

Reconstitution of the electron transport chain from NADH to cytochrome b_5 is achieved upon incorporation of both NADH–cytochrome b_5 reductase and cytochrome b_5 into the same liposomes (Enoch *et al.*, 1977). Under these conditions, the activity of the reductase approaches its maximal rate. Maximal rate is defined as the rate assayed with ferricyanide as electron acceptor instead of cytochrome b_5. The rate of reaction assayed with saturating amounts of either solubilized cytochrome b_5 or heme polypeptide as electron acceptors are only 13 and 4.5% respectively of the maximal activity. Moreover, the oxidation of reductase incorporated into liposomes by saturating amounts of cytochrome b_5 incorporated into other liposomes is only 0.1% of the maximal rate. Since the alternative substrates were present in saturating amounts, the maximal rate obtained by incorporation of both proteins into the same liposomes cannot be explained by a simple concentrative effect. One has to assume that binding both proteins to the same membrane induces correct relative orientation of the catalytic moieties of the two proteins. The catalytic interaction of the reductase and cytochrome b_5 incorporated into synthetic phosphatidylcholine liposomes occurs only above the melting temperature of the lipids. This further stresses the role of lateral protein diffusion in this electron transport chain.

Since the catalytic interaction between the reductase and cytochrome b_5 is highly dependent on the two proteins being in the same vesicle, one can use this as an assay for the transfer of reductase from one vesicle to another. Enoch *et al.* (1977) have incubated cytochrome b_5 proteoliposomes with NADH–cytochrome b_5 reductase proteoliposomes. Initially, the reduction of the cytochrome by the reductase was very sluggish but within two hours the activity of the reductase increased and reached 40–70% of the maximal activity expected. This was interpreted as transfer of reductase molecules from their original liposomes to cytochrome b_5 proteoliposomes. The transfer is maximal at the melting temperature of the liposomes and does not occur below this temperature. The release of reductase from the proteoliposomes is thus dependent on membrane fluidity and is enhanced by disturbances in bilayer structure occurring at the melting temperature.

The binding experiments of cytochrome b_5 and NADH cytochrome b_5 reductase to hepatic microsomes and liposomes have suggested a mechanism for the interaction of amphipatic electron carriers in the endoplasmic reticulum. The manipulation of membrane composition allowed elegant experiments, proving the pivot role of lateral diffusion in protein–protein interaction. The interaction of these amphipathic proteins with membranes has been well characterized but, as will be shown later in this review, this does not seem to apply to reconstitution of integral proteins spanning the membrane.

2.3 CYTOCHROME *c* OXIDASE

Cytochrome *c* oxidase, the terminal member of the mitochondrial electron transport chain catalyzes the simultaneous oxidation of cytochrome *c* with molecular oxygen and the extrusion of protons from the mitochondrial matrix across the inner mitochondrial membrane (Hinkle *et al.*, 1972). In the present chapter we shall describe two levels of reconstitution of cytochrome oxidase functions, the first involves the restoration of cytochrome *c* oxidation capacity to a lipid-depleted preparation of the enzyme and the second concerns the reconstitution of proton translocation.

Cytochrome oxidase is a complex molecule containing two hemes (a and a_3), two copper ions and one or more copies of seven or eight polypeptides (Yonetani, 1961; Kuboyama *et al.*, 1972; Mason *et al.*, 1973; Downer *et al.*, 1976). The isolated enzyme aggregates spontaneously into membraneous structures. Freeze-fracture studies of such membranes revealed prolate spheroids of 135 Å x 68 Å with the long axis extended through the membrane (Hayashi *et al.*, 1972; Vail and Riley, 1974; Henderson *et al.*, 1977). The spheroid seems to correspond to a dimer of the enzyme with a molecular weight of 244 000.

Experiments with bovine heart mitochondria have indicated that cytochrome oxidase spans the mitochondrial inner membrane asymmetrically: a_3, the moiety which interacts with ligands such as carbon monoxide or azide, is situated on the matrix side of the inner membrane, whereas a, the moiety which interacts with cytochrome c is on the outside (Schneider *et al.*, 1972; Palmieri and Klingenberg *et al.*, 1975a). In order to elucidate the arrangement of the various subunits across the membrane, mitochondria and particles in which the 'sidedness' of the inner membrane has been reversed (submitochondrial particles) were exposed to the membrane impermeant surface probe *p*-diazobensene-sulfonic acid (Eytan *et al.*, 1975a; Eytan and Broza, 1978a). In the mitochondria, four subunits and especially subunits II and III were exposed to the label whereas in the inverted particles a different subunit, not identical with either of the others was labeled. Two subunits were not labeled at all and are probably situated in the interior of the membrane. Subunit III of yeast cytochrome oxidase participates in binding cytochrome *c*. Birchmeier and Schatz (1976) have covalently coupled cytochrome *c* to cytochrome oxidase via subunit III; the bound cytochrome *c* serving as a substrate. The asymmetric arrangement of cytochrome *c* oxidase across the membrane fits its dual function, oxidation of cytochrome *c* and generation of proton motive force that can be utilized for ATP synthesis (Mitchell, 1966). Mitchell postulated that the proton translocation is a consequence of the asymmetric translocation of electrons through the enzyme and not actual proton transport. Cytochrome oxidase is assayed *in situ* with ascorbate as reductant. Under these conditions, protons are consumed on the inner side when water is formed from protons and oxygen, while protons are liberated artificially on the external side during oxidation of ascorbate (Hauška *et al.*, 1977). The enzyme translocates electrons from the externally located cytochrome *c* to its internally located ligand binding site. Recently, Wikström (1977), who assayed cytochrome oxidase with ferrocyanide as electron donor, suggested a different enzyme mechanism involving the existence of a proton pump separate from electron translocation.

2.3.1 Reactivation of lipid-depleted cytochrome oxidase

Cytochrome oxidase may be isolated from inner mitochondrial membranes by a variety of procedures involving extraction with non-denaturing detergents, salt precipitations and further purification by chromatography or gradient sedimentation (Yonetani, 1961; Kuboyama *et al.*, 1972; Yu *et al.*, 1973). The isolated enzyme contains variable amounts of bound phospholipids ranging from 0.1–20% of the enzyme's weight. It has been reported that cardiolipin present in the inner mitochondrial membrane is preferentially bound and isolated with cytochrome exidase (Fleischer *et al.*, 1962; Awasthi *et al.*, 1971). Enzyme preparations containing 20% or more phospholipids exhibit maximal cytochrome *c* oxidation (Yu *et al.*, 1975; Eytan and Broza, 1978a). The enzyme may be stripped of bound phospholipids without irreversibly denaturing it by phospholipases, exchanging the bound lipids with mild detergents, or by careful extractions with organic solvents (Fleischer *et al.*, 1962; Brierly and Merola, 1962; Awasthi *et al.*, 1971; Yu *et al.*, 1975; Yu and Yu, 1976; Robinson, 1977). The lipid-depleted enzyme is inactive. Yu and Yu (1976) have shown that, in this enzyme, electron transport from heme *a* to heme a_3 is inhibited. The enzyme may be re-activated by a variety of phospholipids, lipids and even with some detergents like Tween 80 or Emasol 1130 (Yu *et al.*, 1975; Robinson and Capaldi, 1977). The amount of phospholipids required for maximal activation (in absence of added detergents) is equivalent to about half the weight of the protein (Yu *et al.*, 1975; Eytan and Broza, 1978a). However, the amount of lipids actually bound to the enzyme equals 20% of the enzyme's weight (Eytan and Broza, 1978a). The amount remains constant over a wide range of lipid to protein ratios in the incubation mixture, and is equal to the amount observed in most preparations of beef heart cytochrome oxidase. Jost *et al.* (1973) have suggested that there is also a boundary of immobilized lipids between the hydrophobic protein and adjacent fluid bilayer in the membrane. They have studied the electron spin resonance spectra of the lipid spin label, 16-doxylstearic acid, introduced into cytochrome oxidase preparations containing varying amounts of attached phospholipids. At phospholipid to protein ratios lower than 0.19 mg of phospholipid per mg of protein, the electron spin resonance spectra indicated that all the phospholipid is highly immobilized. At higher phospholipid to protein ratios, an additional component was observed corresponding presumably to a phospholipid fraction free to float in the lipid bilayer. Calculations led the authors to suggest that the amount of tightly bound phospholipids corresponds to a single layer of phospholipids surrounding the protein. This lipid fraction was termed annular lipids and seems a general feature of integral membrane proteins (Breulet and McConnell, 1976; Singer and Nicolson, 1972; Warren *et al.*, 1974a,b, 1975; Heron *et al.*, 1977; Thilo *et al.*, 1977). All integral membrane enzymes appear to need the annulus of lipids for activity. The active conformation of these proteins depends on a suitable lipid micro-environment. The specificity of the different enzymes for the annular lipids varies from β-hydroxybutyrate dehydrogenase, which exhibits absolute requirement for phosphatidylcholine with unsaturated fatty acyl groups (Grover *et al.*, 1975), to cytochrome oxidase that may be activated by almost any lipid.

Warren *et al.* (1974a, b, 1975) studied lipid–protein interactions of Ca^{2+}-ATPase from rabbit back muscles. This integral enzyme protein catalyzes *in situ* an ATP-dependent Ca^{2+} transport into the sarcoplasmic reticulum. In the solubilized state, the enzyme catalyzes a Ca^{2+}- and Mg^{2+}-dependent ATP hydrolysis. This activity depends on a proper micro-environment provided by the annular phospholipids (Warren *et al.*, 1975). Removal of bound phospholipids either by organic solvents or detergents leads to partial or full inactivation of the enzyme. However, restoration of the activity is partial and variable, probably due to some irreversible denaturation (Dean and Tanford, 1977; Knowles *et al.*, 1976; Korenbrot, 1977). Warren *et al.* (1974b) have developed a procedure for exchanging the lipids bound to the enzyme with controlled synthetic or purified phospholipids. The enzyme was equilibrated with added excess phospholipids in presence of cholate. This procedure avoids the deleterious effects of lipid depletion and has become a standard procedure for controlling the lipid environment of membrane proteins.

2.3.2 Reconstitution of proton translocation

The proton pumping capacity of cytochrome oxidase has been reconstituted in two lipid model systems: liposomes and planar lipid bilayer.

Most active cytochrome oxidase preparations are suitable for reconstitution of proton translocation. However, the detergents Triton X 100 and deoxycholate used in some of the solubilization procedures of cytochrome oxidase seem to interfere with subsequent reconstitution (Carroll and Racker, 1977). Even cholate, the most commonly used detergent in cytochrome oxidase purification, causes partial denaturation of the enzyme. Carroll and Racker (1977) refractionated seemingly pure cytochrome oxidase preparations and obtained various fractions with similar purity and specific activity. However, only a narrow range of these fractions were suitable for reconstitution of proton translocation. This raises the possibility that the requirements for proton translocation are more stringent than those for cytochrome *c* oxidation and that some cytochrome oxidase preparations contain substances, maybe ionophores, that interfere with subsequent reconstitution.

Hinkle *et al.* (1972) were the first to reconstitute proton-translocating cytochrome oxidase proteoliposomes. Soy bean phospholipids and purifed cytochrome oxidase were dissolved with cholate. Subsequently, proteoliposomes active in proton translocation were formed upon slow dialysis of the detergent. Experiments with radioactive [^{14}C] cholate revealed that the half-time of cholate removal was 75 minutes (Kagawa *et al.*, 1973a,b). Residual cholate remained firmly bound to the vesicles and specifically to the phosphatidylcholine population. Since then, it has become evident that reconstitution of cytochrome oxidase is not such a magic art as it seemed at first. Active proteoliposomes are formed also when the detergent is rapidly removed by chromatography on Sephadex or even when the detergent is not removed at all but simply diluted in the assay mixture (Racker, 1972b; Racker *et al.*, 1975b). A method which proved very successful in reconstitution of other membrane proteins is the sonication of the protein with liposomes (Racker, 1973). However, this procedure is not suitable for reconstitution of cytochrome oxidase as the latter is denatured by even a brief exposure to sonication.

Recently, it has been demonstrated that cytochrome oxidase may be incorporated directly into preformed liposomes (Eytan *et al.*, 1975b, 1976a). This incorporation occurs presumably by fusion of the liposomes with the isolated enzyme (Eytan, 1978). The direct incorporation is highly dependent on the liposomes' composition. It is facilitated by the presence of either low concentrations of detergents (Eytan *et al.*, 1975b), or by acidic phospholipids (e.g. cardiolipin, phosphatidylserine and phosphatidylinositol) (Eytan *et al.*, 1976b). The direct incorporation of cytochrome oxidase into preformed liposomes containing acidic phospholipids is most suitable for studies of the mechanism of proteoliposomes reconsitution, since it does not involve the addition of detergents or sonication. Experiments with ^{86}Rb trapped inside the liposomes have demonstrated that the protein was inserted with no increase in the leakage rate of the liposomes content.

Much effort has been invested lately in incorporation of cytochrome oxidase into planar lipid bilayers. Limited success has been achieved by initially forming a complex of the enzyme and phospholipids with Ca^{2+} ions. This lipoprotein was soluble in hexane and as such could be applied to an aperture in a teflon partition separating two aqueous phases. The hexane was drawn to the hydrophobic teflon and a thin proteolipid membrane was formed across the aperture allowing electrical measurments of the proton pump. However, this procedure is tedious and rather difficult to reproduce. The technical problems involved in introducing a protein into planar lipid model system were recently circumvented by the group of Skulachev (Drachev *et al.*, 1974, 1976a,b). They incorporated cytochrome oxidase, among other proteins into liposomes and subsequently in presence of high Ca^{2+} concentrations, the proteoliposomes were absorbed on a planar lipid bilayer. This system allowed studies of the electrical properties of cytochrome oxidase-catalyzed proton translocation.

2.3.3 Assays of cytochrome oxidase-catalyzed proton translocation

Reconstituted cytochrome oxidase proteoliposomes extrude protons (Hinkle *et al.*, 1972). This has been demonstrated by suspending the proteoliposomes in an anaerobic solution containing reduced cytochrome *c* and the reductant dihydronaphtoquinone. Upon introduction of oxygen there occured acidification of the solution which was recorded with a pH meter. The stoichiometry exhibited by the system was 2 protons extruded per atom oxygen consumed (Hinkle *et al.*, 1972). Since the assays were performed in presence of valinomycin which renders the proteoliposomes permeable to potassium ions, it has been possible to demonstrate that two potassium ions were pumped into the liposomes per atom oxygen consumed, thus neutralizing the electrical potential generated by the proton pump. Proton translocation was unidirectional since only enzyme molecules oriented with their cytochrome *c* binding moiety exposed could interact with the externally added cytochrome *c*.

A simpler, although indirect, assay for proton translocation is the measurement of respiratory control of cytochrome *c* oxidation (Racker, 1972b, Hinkle *et al.*, 1972). As the oxidation of cytochrome *c* is coupled to proton extrusion from the proteoliposomes after a short initial period, its rate will be restricted by the rate protons can leak back into the proteoliposomes. As liposomes are relatively impermeable to protons, the activity of properly incorporated cytochrome oxidase will be severely inhibited. This inhibition may be relieved by rendering the liposomes permeable to protons with uncouplers or by disrupting the liposomes structure with mild detergents. Respiratory control is defined as the ratio of maximal activity, assayed as oxygen consumption in presence of reduced cytochrome *c* and uncouplers, to the activity assayed in absence of uncouplers. This ratio is often used as a parameter for functional reconsitution (Eytan *et al.*, 1975b, 1976a,b; Carroll and Racker, 1977).

Skulachev *et al.* (Drachev *et al.*, 1974, 1976a,b) and Montal (1974a,b) have introduced new and elegant approaches to the assay of proton translocation by cytochrome oxidase vesicles. For example, they prepared cytochrome oxidase proteoliposomes with cytochrome *c* trapped inside by including the latter in the dialysis mixture. External cytochrome *c* was washed away. Since ascorbate and other water-soluble reductants do not penetrate liposomes, the internal cytochrome *c* had to be reduced with the lipid-soluble reductant phenazine methosulfate. Under these conditions, introduction of oxygen induced proton uptake instead of extrusion, as only enzyme molecules oriented with their cytochrome *c* binding sites exposed to the inner volume were functional. Proton uptake was assayed indirectly by introduction of a lipid-soluble anion phenyl-carbaundecaborane (PCB^-) which is co-transported with the protons pumped into the vesicles. Skulachev *et al.* (Drachev *et al.*, 1974, 1976a,b) assayed the PCB^- translocation using a planar lipid bilayer as a sensitive electrode and indeed could detect a substantial proton translocation into the vesicles which was abolished by uncouplers and cyanide. Furthermore, they have introduced cytochrome oxidase into planar lipid membrane either directly as a decane-soluble lipoprotein or indirectly by adsorbing cytochrome oxidase proteoliposomes onto the planar lipid bilayer. In the first case, they observed that upon addition of cytochrome *c* and ascorbate to one of the two compartments separated by the cytochrome oxidase containing planar membrane an electric potential was formed (plus on the cytochrome *c* side of the membrane). Maximal values of transmembraneous potential differences obtained by this method were 40 mV. In similar assays, with cytochrome oxidase proteoliposomes adsorbed to the planar membrane, an electric potential difference of up to 100 mV was generated and a current of 1×10^{-11} A was measured.

2.3.4 Optimal phospholipid to protein ratio for reconstitution

Reconstitution of cytochrome oxidase by the cholate dialysis, and direct incorporation procedures requires a 20-fold excess of phospholipids over protein (Carroll and Racker, 1977; Racker, 1972b; Eytan *et al.*, 1976a). The incorporation of the enzyme into liposomes containing low detergent requires a phospholipid to protein ratio of 200 (Eytan *et al.*, 1975b). High lipid to protein ratios were required also for reconstitution of other proteins. A few representative examples are the Na^+ K^+ATPase from electric eel (Racker *et al.*, 1975b; Goldin and Tong, 1974; Hilden *et al.*, 1974) and the Ca^{2+}ATPase from rabbit back muscle (Racker, 1972a) which required a ratio of 20:1, the oligomycin-sensitive ATPase (Kagawa and Racker, 1971; Kagawa *et al.*, 1973a,b) and QH_2-cytochrome *c* reductase (Leung and Hinkle, 1975; Eytan *et al.*, 1976a) from inner mitochondrial membrane which required a 2–5-fold excess of phospholipids and the extreme case of band 3 of red blood cells which required a 29 000-fold excess of phospholipids over protein (Ross and McConnell, 1977). The high lipid to protein ratios required for reconstitution are puzzling since the lipid to protein ratios observed in biological membranes vary between 0.3:1 in complex membranes like the inner mitochondrial membrane, and a ratio of 1:1 in a relatively simple membrane like the myelin sheath.

Calculations based on the minimal molecular weight of cytochrome oxidase and the average molecular weight of a liposome indicate that at a lipid to protein ratio of 20:1, each liposome contains on the average one molecule of enzyme. The high lipid to protein ratios would then be explained if one assumes that each liposome can accomodate only one enzyme molecule. This assumption seems far-fetched and indeed Carroll and Racker (1977) who have studied the distribution of the incorporated enzyme found it to be non-random. They have incorporated cytochrome *c* oxidase into preformed vesicles and analyzed them by centrifugation into a Ficoll gradient. Two liposome subpopulations were obtained, a proteoliposome fraction containing all the enzyme but only 5–10% of the phospholipids and a protein-free liposome fraction containing the rest of the liposomes. The latter fraction is not suitable for incorporation. When cytochrome oxidase is reconstituted at a lipid to protein ratio lower than 20 the excess protein is not incorporated into these liposomes but remains as a protein–annular lipids complex (Carroll and Racker, 1977; Eytan and Broza, 1978a). Moreover, incubation of the protein-free liposomes isolated from the Ficoll gradient with additional amounts of enzyme does not lead to functional incorporation. Carroll and Racker have shown that the protein-free liposomes do not lack a chemical component required for incorporation since the lipids of these liposomes could be extracted and reused to form liposomes. The new liposomes were functional in incorporation.

Recent studies have indicated that the characterisitic distinguishing the liposomes suitable for incorporation from the rest is their size. Eytan and Broza (1978b) fractionated liposomes containing acidic phospholipids according to their size on a Sepharose column. The various fractions were incubated with cytochrome oxidase at a constant phospholipid to protein ratio. Only a narrow range of fractions containing liposomes of approximately 220 Å diameter were suitable for incorporation. On the other hand, cytochrome oxidase had been incorporated into non-fractionated liposomes and the resulting proteoliposomes were fractionated as above. The enzyme was associated with liposomes of a 220–350 Å diameter. Thus, it seems that upon incorporation of cytochrome oxidase into liposomes, the latter grow from a diameter of 220 Å to a diameter of 220–350 Å. The reconstitution of protein into only a subpopulation of liposomes is not limited to direct incorporation into preformed liposomes. The high lipid to protein ratios optimal for reconstitution by cholate dialysis and cholate dilution may point to the possibility that, in these cases too, the enzyme is not uniformly distributed among all liposomes. It has been shown that following cholate dialysis cytochrome oxidase is reconsituted only in a liposome subpopulation with a diameter of 220 Å–350 Å.

Recently, it has been demonstrated that upon rapid dialysis of cholate, cytochrome oxidase (Carroll and Racker, 1977) and Na^+ K^+ATPase from renal medulla (Goldin, 1977) were successfully reconstituted at relatively low lipid to protein ratios. The Na^+ K^+ATPase has been shown to be randomly distributed among the liposomes with one molecule of enzyme present on the average in each liposome. It is tempting to speculate that under standard cholate dialysis conditions, liposomes are formed first and only later does the enzyme become insoluble and attached to the liposomes' subpopulation suitable for incorporation. On the other hand, under rapid dialysis conditions the enzyme is incorporated before the liposomes are completely formed and thus it is incorporated randomly into all liposomes.

The subpopulation of liposomes suitable for reconstitution is characterized by the small diameter of the liposomes. These liposomes are distinguished by the high curvature of their membranes and, as a consequence by asymmetry in the distribution of phospholipids across the membranes. Further experiments are needed to define the characteristic essential for reconstitution. The curvature induces an increase in surface tension of the membranes which has been shown to facilitate fusion of liposomes containing acidic phospholipids (Miller and Racker, 1976; Miller *et al.*, 1976). As will be discussed later, the proteins are incorporated unidirectionally into the liposomes. The asymmetry of the phospholipid bilayer might participate in the orientation of the proteins and provision of suitable microenvironment for their functions.

2.3.5 Orientation of incorporated proteins

Reconstitution by cholate dialysis or dilution may yield mixed proteoliposomes populations with respect to the orientation of incorporated membrane proteins. For example, roughly 40% of the activity of oligomycin-sensitive ATPase vesicles prepared by the cholate dilution procedure could be detected only in the presence of detergent (Eytan *et al.*, 1976a). In contrast, all the activity of oligomycin-sensitive

ATPase reconstituted by direct incorporation into preformed liposomes could be assayed even in the absence of detergent. This implies that in the latter case, most, if not all of the enzyme was oriented unidirectionally, with its hydrolytic site exposed to the medium. Random orientation of protein reconstituted by the cholate dialysis procedure has been reported also for the Na^+ K^+ATPase from renal medulla (Goldin, 1977).

Cytochrome oxidase, on the other hand, is oriented unidirectionally, whether reconstituted by the cholate dialysis procedure or by direct incorporation into proteoliposomes (Racker, 1972b; Carroll and Racker, 1977; Eytan *et al.*, 1976a). Since cytochrome *c* does not penetrate liposomes, the fraction of cytochrome oxidase oriented with its cytochrome *c* binding site exposed can be assessed by comparing the enzyme's activity in presence of uncouplers to that assayed in presence of detergents. Of several detergents tested, the nonionic detergent, Tween 80, has been found the most suitable as it disrupts liposomes without inhibiting the enzyme. The uncoupled activity of cytochrome oxidase proteoliposomes is similar to that assayed in presence of Tween 80 and to that of purified enzyme containing a complete anullus of lipids. An alternative procedure, useful in determination of cytochrome oxidase orientation in proteoliposomes, is surface labeling of exposed subunits (Eytan *et al.*, 1975a). Subunits II and III were the most exposed to surface labeling with [^{35}S] diazonium benzene sulfonate, both in the proteoliposomes and, as mentioned above, in intact mitochondria (Eytan and Broza, 1978a). Similar conclusions were reached by Kornblatt *et al.* (1975). They have prepared cytochrome oxidase, specifically labeled at subunit III with a spin label derivative of *N*-ethylmaleimide. Upon reconstitution of this enzyme into proteoliposomes, all the spin label remained exposed to reduction by externally added ascorbate. Since ascorbate does not penetrate closed proteoliposomes, they too have concluded that cytochrome oxidase is oriented unidirectionally with its subunit III exposed to the external medium. Thus, both functional assays and direct chemical analysis indicate that the orientation of cytochrome oxidase in proteoliposomes corresponds to its orientation in mitochondria.

In contradistinction, cytochrome oxidase reconstituted by cholate dialysis in presence of high concentrations of ferricytochrome *c*, is embedded symmetrically in the membrane (Drachev *et al.*, 1974; Carroll and Racker, 1977). In this case, half the enzyme is embedded, as in the mitochondria, and can be assayed with external ascorbate and cytochrome *c* while the other half is oriented with its cytochrome *c* binding site exposed to the inner medium of the liposomes. The activity measured as oxygen consumption of the latter enzyme fraction can be assayed only in presence of a permeable reductant such as phenazine methosulfate capable of reducing the internally trapped cytochrome *c*. Skulachev *et al.* (Drachev *et al.*, 1974) have assayed proton translocation in these proteoliposomes. In the presence of external cytochrome *c* and impermeant reductant the proteoliposomes translocate protons outwards. Proton translocation inwards, catalyzed by the appropriate enzyme fraction could be assayed only after careful washing of external cytochrome *c* and in presence of phenazine methosulfate. The cause for scrambling of cytochrome oxidase in the proteoliposomes is the presence of oxidized cytochrome *c* during the reconstitution. Reconstitution in the presence of reduced cytochrome *c* (i.e. under anaerobic conditions with ascorbate) results in unidirectionally oriented enzyme with its cytochrome *c* binding site exposed. Oxidized cytochrome *c* seems to bind to cytochrome oxidase and thereby impair its capacity for unidirectional orientation.

Racker and Kandrach (1973) have demonstrated the major role of asymmetric orientation in membrane function. They have reconstituted proteoliposomes containing both cytochrome oxidase and oligomycin-sensitive ATPase in the presence of cytochrome *c*. ATP was generated in these particles through the activity of the ATPase and cytochrome oxidase molecules oriented in the submitochondrial configuration. The phosphorylation was absolutely dependent on complete inhibition of the cytochrome oxidase fraction oriented right side out. The inhibition was achieved by washing away external cytochrome *c* and further inhibiting the cytochrome oxidase with the impermeant competitive inhibitor polylysine.

The mechanism of anisotropic assembly of proteoliposomes is best studied in direct incorporation of proteins into preformed liposomes, where detergents and some disruption of the liposomes are avoided. The unidirectional incorporation of cytochrome oxidase is probably an outcome of an asymmetric feature of the isolated enzyme, either different hydrophobicity of the two poles of the elongated enzyme molecule or different electric charge. As the cytochrome *c* binding site of the enzyme is acidic and the overall charge of the enzyme is neutral, the opposite pole of the enzyme is possibly positive. A feasible mechanism of asymmetric orientation is initial electrostatic binding of the basic moiety of the enzyme with the acidic liposomes. This was tested by incubating cytochrome oxidase with liposomes rendered positive with the basic synthetic phospholipid stearylamine. In these liposomes, the activity of cytochrome oxidase is strongly inhibited, a feature that could have indicated reverse orientation of the enzyme. However, no increase in activity was observed when cytochrome *c* has been trapped inside the liposomes and phenazine methosulfate included in the assay mixture. Furthermore, chemical labeling of the proteoliposomes with *p*-diazonium benzene sulfonate resulted in tagging of subunits II and III (Eytan and Broza, 1978a). As mentioned above, these subunits are part of the cytochrome *c* binding moiety. The inhibition of the enzyme is probably due to reversible denaturation which is evident in the exposure of the usually deeply buried subunit I. Thus, electrostatic interaction does not seem to play a role in the asymmetric incorporation of cytochrome oxidase and one is left with hydrophobic interaction as the main orientation factor. Chemical labeling experiments described above have shown that at least four subunits are exposed on the cytochrome *c* binding moiety of the enzyme while only one subunit is exposed on the opposite pole. Similarly, the oligomycin-sensitive ATPase is incorporated with the hydrophobic moiety buried in the membrane and the hydrophilic peripherial hydrolytic moiety (F_1) exposed to the medium. Similar conclusions were reached by Wickner (1977) who studied the incorporation of a viral polypeptide into liposomes. This author also eliminated electrostatic binding as the cause of the unidirectional orientation of the protein in the membrane, and suggested hydrophobic interaction as the decisive factor. In contradistinction, the orientation of bacteriorhodopsin, reconstituted by brief sonication, is controlled by electrostatic interaction. Bacteriorhodopsin, usually reconsituted 'inside out' compared to its orientation in the bacterium, is incorporated 'right side out' when sonicated with liposomes rendered neutral in acidic pH (Happe *et al.*, 1977).

Another possible factor directing the anisotropic orientation of membrane proteins is the asymmetric distribution of the lipids (Bergelson and Barsukov, 1977). The phospholipid composition of the two monolayers comprising the liposomes is significantly different in small liposomes. This asymmetry might have induced asymmetry of the proteins. A possible hint to the importance of asymmetry in reconstitution is the experiment reported by Kagawa *et al.* (1973b). They have reconstituted proteoliposomes containing phosphatidylethanolamine, cardiolipin and the oligomycin-sensitive ATPase. The proteoliposomes were inactive in ^{32}Pi-ATP exchange (see below) but could be activated by transfer of phosphatidylcholine with phosphatidylcholine transfer protein from bovine heart. This enzyme has recently been shown to transfer phospholipids exclusively to the external monolayer of liposomes and thus it induces asymmetry in liposomes formed in its presence. This asymmetry might have been important in orientation and function of the oligomycin-sensitive ATPase.

2.3.6 Sequential incorporation of proteins into liposomes

One of the intriguing features of *in vivo* insertion of proteins into membranes is the selectivity of the process. Proteins are incorporated into the appropriate membrane in the presence of other membranes. A possible approach to the mechanism of the selection is studies of sequential incorporation of membrane proteins into liposomes. The direct incorporation of proteins into preformed liposomes (Eytan *et al.*, 1976a,b) is most suitable for this purpose as it does not involve the use of either detergents or sonication. The most striking effects were observed on incorporating cytochrome oxidase and the hydrophobic moiety of the oligomycin-sensitive ATPase into the same liposomes (Eytan and Racker, 1977). The presence of the two proteins in the same vesicle could be conveniently assayed, since the hydrophobic protein fraction serves as a proton channel and abolishes respiratory control of the incorporated cytochrome oxidase (Racker, 1972b). The determining factor in the effectiveness of the hydrophobic protein is not its absolute concentration but the ratio of this fraction to cytochrome oxidase vesicles (Eytan and Racker, 1977). Thus, in the presence of increasing amounts of cytochrome oxidase vesicles, a given concentration of hydrophobic proteins became less effective for uncoupling. Does cytochrome oxidase, already incorporated into liposomes, affect the subsequent incorporation of hydrophobic proteins of mitochondrial ATPase? To answer this question, a fixed amount of hydrophobic protein fraction was incubated with liposome mixtures containing a fixed amount of cytochrome oxidase vesicles and increasing amounts of protein-free liposomes. If the hydrophobic proteins are randomly incorporated into both types of liposomes, the presence of excess of protein-free liposomes should decrease the uncoupling (i.e. raise the respiratory control ratio) of the cytochrome oxidase vesicles. However, this was not the case; the degree of uncoupling remained constant despite the presence of excess protein-free liposomes. All the added hydrophobic protein had therefore been incorporated only into the cytochrome oxidase vesicles. This result is all the more surprising since both liposome populations had the same phospholipid composition. Essentially, the same result was obtained when the cytochrome oxidase vesicles were prepared from an enzyme that had been almost totally depleted of lipids and detergents thus excluding the trivial explanation that the cytochrome oxidase introduced detergents or extra phospholipids into the liposomes.

The phenomenon is intriguing as it is not a general phenomenon but dependent on the protein species involved. Cytochrome oxidase was incubated with a series of liposome mixtures consisting of a fixed amount of liposomes with incorporated hydrophobic proteins and increasing amounts of protein-free liposomes. Preferential incorporation of the cytochrome oxidase into the liposomes containing hydrophobic protein should have resulted in particles lacking respiratory control. Conversely, random incorporation of cytochrome oxidase into the two types of liposomes should have yielded respiratory control values that increased with increasing amounts of protein-free liposomes. The observed results were even higher than those expected for a random distribution of cytochrome oxidase. Thus, cytochrome oxidase was incorporated preferentially into the protein-free liposomes. Depending on the system investigated, the presence of incorporated proteins in liposomes may therefore stimulate or inhibit the subsequent incorporation of a second protein. Further efforts will be required to learn whether these effects reflect direct protein–protein interaction or a modification of the liposomes by the incorporated proteins. The selectivity exhibited in sequential incorporation of membrane proteins into liposomes might also suggest that *in vivo* the information needed for correct incorporation resides in the membrane proteins proper. It has been shown that subtle changes in the proteins already incorporated into liposomes appear to alter markedly the affinity of other proteins towards the same liposomes (Eytan and Racker, 1977). For example, partial reduction of cytochrome oxidase incorporated into liposomes, or binding of cytochrome *c*, abolishes the affinity of the hydrophobic protein fraction towards the proteoliposomes. This might point to a possible mechanism for modulating the incorporation of membrane proteins. Biological examples of such delicate modulations are the regulation of mitochondrial biogenesis by oxygen (Slonimski, 1953; Criddle and Schatz, 1969), glucose (Mahler, 1973) or heme (Gollub *et al.*, 1974) and the regulation of chloroplast membrane formation by the chlorophyll synthesizing system (Ohad, 1975). However, in making an analogy with membrane biogenesis *in vivo*, one has to keep in mind that the inner mitochondrial membrane, for instance, is not formed by insertion of whole complexes such as cytochrome oxidase or the oligomycin-sensitive ATPase but by incorporation of individual or even nascent chains (Schatz and Mason, 1974; Kellems and Butow, 1972).

In summary, despite the reservation mentioned above about the applicability of data obtained from proteoliposome reconsitution to membrane biogenesis *in vivo* we feel that some general rules deduced from the artificial model system can be applied to the biological problem. (a) The acidic phospholipids present ubiquitously in biological membranes seem to promote incorporation of newly formed proteins or polypeptides. (b) The proteins already present in the membranes may increase or decrease the affinity of a newly formed protein towards the membrane. (c) Subtle changes in the proteins such as oxidation-reductions appear to alter markedly the complex mutual influence between the proteins with respect to their incorporation into membranes.

2.3.7 The role of membrane fluidity in reconstitution of proteoliposomes

Fully saturated synthetic phospholipids inhibit the reconstitution of some proteins such as cytochrome oxidase (Eytan *et al.*, 1976a,b; Eytan and Broza, 1978a), oligomycin-sensitive ATPase (Kagawa, 1973a) and β-hydroxybutyrate dehydrogenase (Grover *et al.*, 1975; Jurtshuk, 1961), whereas they do not seem to interfere with reconsitution of other proteins, e.g. bacteriorhodopsin (Racker and Hinkle, 1974), Ca^{2+}-ATPase (Warren *et al.*, 1974a,b) and QH_2-cytochrome *c* reductase (Eytan *et al.*, 1976a). These phospholipids reduce the fluidity of phospholipid bilayers and the inhibition, where it occurs, indicates probably that the enzyme requires a certain minimal fluidity for its function. The inhibition is not caused by an increased leakiness of the liposomes since experiments with ^{86}Rb-loaded liposomes revealed no effect of saturated phospholipids on rubidium release (Eytan *et al.*, 1976a).

Cytochrome oxidase incubated with vesicles prepared with fully saturated phospholipids, such as dipalmitoylphosphatidylcholine or dimyristoylphosphatidylcholine did not exhibit respiratory control. An exception to this rule was observed with saturated phospholipids containing short fatty acyl groups such as dilauroylphosphatidylcholine. The latter phospholipid differs from the first ones in its lower melting temperature and consequently more fluid nature. Thus, the exception to the rule seems to prove the rule. Moreover, short chain alcohols, such as pentanol, which are known to increase the fluidity of membranes (Dawson *et al.*, 1976), counteract the inhibition by saturated phospholipids. Cytochrome oxidase reconstitution is inhibited by cholesterol as well. Inclusion of cholesterol in liposomes leads to a more rigid and compact structure (Feinstein *et al.*, 1975).

An interesting feature of the inhibition exerted by saturated phospholipids is that the degree of inhibition is not correlated to the overall concentration of saturated fatty acyl groups but to the concentration of fully saturated phospholipids molecules. This has been shown (Eytan *et al.*, 1976b) by incubation of cytochrome oxidase with liposomes containing varying amounts of fully saturated synthetic phosphatidylcholine and either egg or mitochondrial phosphatidylcholine. Despite the marked difference in saturation between the two natural phosphatidylcholines, the amount of saturated phosphatidylcholine needed to inhibit functional incorporation was similar in both cases. The saturated phospholipids seem to interfere with the incorporation proper of cytochrome oxidase into liposomes and not with the function of already incorporated enzyme. This has been shown by fusing active proteoliposomes containing the enzyme and natural lipids with liposomes containing saturated synthetic phospholipids. The resultant proteoliposomes were functionally active despite the synthetic phospholipids present in them. Moreover, fusion of functionally defective cytochrome oxidase vesicles containing saturated phospholipids with excess liposomes containing natural lipids did not correct the defect in cytochrome oxidase function. The detailed mechanism of the inhibition remains to be clarified. The saturated phospholipids do not interfere with initial binding of the enzyme to the liposomes (Eytan and Broza, 1978a). Cytochrome oxidase had been incubated with liposomes containing fully saturated phospholipids and the resultant mixture was centrifuged into a Ficoll gradient. The enzyme, although associated with liposomes, exhibited no respiratory control. Thus it seems that the fully saturated lipid inhibited the penetration of the enzyme into the liposomes membranes but not the initial binding of the enzyme to the liposomes. An alternative experiment leading to the same conclusion is incubation of cytochrome oxidase with a mixture of liposomes containing saturated phospholipids and liposomes containing only natural phospholipids (Eytan and Racker, 1977). The enzyme developed an intermediate respiratory control, indicating that a fraction of the enzyme was bound to the inhibitory liposomes resulting in defective incorporation. On the other hand, initial fusion of the two liposome population resulting in homogeneous liposomes population containing saturated phospholipids and subsequent incubation with the enzyme resulted in complete inhibition of respiratory control.

In summary, fully saturated phospholipids seem to interfere with the proper incorporation of cytochrome oxidase into the liposomes but not with the initial binding of the enzyme to the liposomes. The initial binding is dependent on the nature of the headgroups of the phospholipids whereas the nature of the fatty acyl group is critical for the correct orientation of the protein in the membrane.

2.4 RECONSTITUTION OF MITOCHONDRIAL TRANSPORT SYSTEMS

The inner mitochondrial membrane contains a large number of membrane proteins. A major group is involved in oxidative phosphorylation. In this overall process electrons are withdrawn from a variety of substrates (oxidation) and transferred via the respiratory chain to oxygen. This stepwise electron transfer is coupled to ATP synthesis. A second group of membrane proteins is involved in the transport of solutes.

P. Mitchell suggested the chemiosmotic theory of oxidative phosphorylation (Mitchell, 1966) and of solute accumulation (Mitchell, 1963). The transfer of electrons through the various complexes of the respiratory chain is coupled to the extrusion of protons. This results in the formation of a proton electrochemical gradient across the inner mitochondrial membrane. This gradient consists of two components, a pH gradient and a membrane potential. The two components are interconvertible. The electrochemical gradient can be utilised by the reversible mitochondrial ATPase complex to drive synthesis of ATP. Alternatively, this proton gradient can drive transport of solutes, by coupling the solute fluxes to proton translocation. The protein complexes able to reversible utilise energy stored in chemical bonds to create proton gradients are defined as primary active transport systems. These include the respiration-driven proton-translocating complexes (Mitchell and Moyle, 1967) such as (a) NADH dehydrogenase (Lawford and Garland, 1972) (b) the cytochrome b–c_1 segment of the respiratory chain (Papa *et al.*, 1975; Lawford and Garland, 1973), (c) cytochrome *c* oxidase (Hinkle *et al.*, 1972) and (d) the energy-dependent pyridine nucleotide transhydrogenase (Rydström, 1977) as well as the proton-translocating ATPase complex (oligomycin sensitive ATPase–F_0–F_1) (Mitchell and Moyle, 1968; Thayer and Hinkle, 1973). Solute translocators utilising ion gradients are defined as secondary active transporter systems. Mitochondrial examples include the Ca^{2+} uptake system (Lehninger, 1970), the phosphate translocator (Chappell and Crofts, 1965) and the adenine nucleotide translocator (Klingenberg and Pfaff, 1966).

The breakthrough of reconstituion by E. Racker and co-workers has been achieved using these mitochondrial transport systems (Kagawa and Racker, 1971; Kagawa *et al.*, 1973a; Racker and Kandrach, 1971, 1973; Ragan and Racker, 1973; Ragan and Hinkle, 1975; Ragan and Widger, 1975; Hinkle *et al.*, 1972; Leung and Hinkle, 1975; Rydström *et al.*, 1975; Serrano *et al.*, 1976). This has resulted in reconstitution of almost all these systems. The examination of the properties of these systems as well as the study of their interactions, have led to an impressive body of evidence which basically supports both the chemiosmotic hypothesis of oxidative phosphorylation (Mitchell, 1966) and that for solute translocation (Mitchell, 1963).

In the present section we will describe reconsitution studies on both primary and secondary transport proteins, located in the inner mitochondrial membrane. We shall refer as well to a few selected examples of analogous transport proteins from other sources, such as the ATPase complex ($TF_0 - TF_1$) from a thermophilic bacterium.

2.4.1 The respiration-driven proton-translocating complexes

These complexes catalyse proton translocation concomitant with electron transfer (Papa, 1976). In intact mitochondria this translocation is outward when electron transfer proceeds toward oxygen (Mitchell and Moyle, 1967) and is opposite in submitochondrial particles prepared by sonication (Hinkle and Horstman, 1971). As described in Section 2.4.2 for the ATPase complex, catalytic activity (here: electron transfer) can proceed in solution, but obviously proton translocation can be observed using vesicular structures only. The three complexes of the respiratory chain, which contain coupling sites I–III have been incorporated in artificial single-walled lipid vesicles. These proteoliposomes catalysed electron transfer-dependent proton translocation (Hinkle *et al.*, 1972; Wikström, 1977; Wikström *et al.*, 1977.

Outward proton translocation was observed in cytochrome oxidase vesicles (Hinckle *et al.*, 1972; Wikström, 1977; Wikström *et al.*, 1977). In these vesicles electron transfer proceeds from ascorbate to oxygen. Proton extrusion was also observed (Leung and Hinkle, 1975; Guerrieri and Nelson, 1975) in vesicles containing the $b{-}c_1$ segment of the respiratory chain (electron flow from reduced ubiquinone via cytochrome *c* to ferricyanide). This polarity is expected since in both cases the impermeant cytochrome *c* was added from the outside (Hinkle *et al.*, 1972; Leung and Hinkle, 1975). On basis of the recovery of electron transfer activity in reconstituted vesicles as compared with the isolated complex, it has been suggested that the orientation of these complexes in reconstituted vesicles is similar to that observed in intact mitochondria.

After reconstitution, an apparent submitochondrial orientation was noted for the NADH dehydrogenase complex [complex I (Ragan and Hinkle, 1975)] and for the transhydrogenase complex (Rydström *et al.*, 1975). Because of the addition of pyridine nucleotides to the outside of the proteoliposomes, only those complexes incorporated in the submitochondrial conformation would be expressed. Whereas with reconstituted complex I vesicles electron transfer-dependent proton translocation was the criteria for reconstitution (Ragan and Hinkle, 1975), for the transhydrogenase this was electron transfer-dependent generation of a membrane potential (Rydström *et al.*, 1975). Recently, the transhydrogenase protein has been purified to homogeneity and an apparent molecular weight of 97 000 for the reconstitutively active protein has been reported (Höjeberg and Rydström, 1977).

The activities of these respiratory complexes was inhibited by the appropriate inhibitors such as rotenone (Ragan and Hinkle, 1975), antimycin (Leung and Hinkle, 1975) and cyanide (Racker and Kandrach, 1973). Moreover, uncoupling conditions enhance the electron transfer activities of all these complexes and abolish the observed proton translocation. This can be explained by the ability of the uncouplers to collapse the proton gradient. On the other hand, proton translocation was enhanced by ionophores which collapse the membrane potential. Therefore this process appears to be electrogenic. The reconstitution studies described in this section have been done with proteoliposomes obtained by cholate dialysis. In addition, successful reconstitution was also achieved using the incorporation method (Eytan *et al.*, 1976a).

2.4.2 The proton-translocating adenosine triphosphatase complex

(a) *The complex from mitochondria*

This complex is involved in the terminal stages of oxidative phosphorylation (Racker, 1970; Harold, 1972). The detergent-solubilised complex catalyses ATP hydrolysis which is inhibited by energy transfer inhibitors such as oligomycin, rutamycin and DCCD (Kagawa and Racker, 1966; Tzagoloff *et al.*, 1968; Kopaczyk *et al.*, 1968; Kagawa and Racker, 1971; Swanljung *et al.*, 1971, 1973; Sadler *et al.*, 1974; Hatefi *et al.*, 1974; Tzagoloff and Meagher, 1971; Ryrie, 1975; Serrano *et al.*, 1976). This activity is dependent on phospholipids, but not on incorporation of this complex in closed single-walled liposomes. In contradistinction, for reactions catalysed by the ATPase which required energy conservation, this incorporation appears to be necessary (Kagawa and Racker, 1971; Serrano *et al.*, 1976).

Such a reaction is the $^{32}P_i$-ATP exchange, which is sensitive to uncouplers of oxidative phosphorylation and energy transfer inhibitors. This activity can be detected in mitochondria and in various submitochondrial particles (Racker, 1970), but not in detergent-solubilised complexes (Kagawa and Racker, 1971; Serrano *et al.*, 1976). When such a detergent-solubilised preparation was properly incorporated into closed single-walled liposomes, $^{32}P_i$-ATP exchange sensitive to both classes of the above mentioned inhibitors was determined (Kagawa and Racker, 1971). This was the first case of functional reconstitution of any membrane transport protein.

Functional mitochondrial ATPase complexes have been isolated, which, upon incorporation, catalysed ATP-dependent proton translocation (Kagawa *et al.*, 1973a,b; Serrano *et al.*, 1976) sensitive to uncouplers and energy transfer inhibitors. Since in all assays of reconstituted ATPase vesicles, ATP was added from the outside, it is clear that only the activity of those complexes incorporated in the submitochondrial conformation is measured.

For many of the above-mentioned experiments in which proton-translocating ATPase has been reconstituted, the cholate dialysis technique was used. Reconstitution techniques such as sonication (Racker, 1973), incorporation (Eytan *et al.*, 1976a) and cholate dilution (Racker and Widger, 1975) have been proven successful for reconstitution. The latter technique has been of great help as an assay for functional ATPase during the purification of detergent-solubilised preparations (Serrano *et al.*, 1976). Active highly purified preparations have been obtained (Serrano *et al.*, 1976; Ryrie, 1975; Hatefi *et al.*, 1974).

The ATPase complex (F_0-F_1) can be dissected into the F_0 and the F_1 portion (Racker, 1970). The relatively hydrophilic F_1 moiety contains the catalytic site. ATP hydrolysis catalysed by F_1 is insensitive to oligomycin rutamycin and DCCD (Racker, 1970). When F_1 is recombined with the hydrophobic F_0 unit, the ATPase activity becomes sensitive to the energy transfer inhibitors. The F_0 portion of the complex appears to contain a proton channel (Hinkle and Horstman, 1971) which can be blocked by the energy transfer inhibitors and by F_1. The proton channel-containing unit has been incorporated into liposomes and has been shown to display energy transfer inhibitor-sensitive proton permeability

(Shchipakin *et al.*, 1976). F_0-containing liposomes, of course do not display any $^{32}P_i$-ATP exchange but, upon addition of F_1 and a polypeptide involved in binding of F_1 to F_0, the oligomycin sensitivity conferral protein [OSCP (Maclennan and Tzagoloff, 1968)], this activity is restored (Serrano *et al.*, 1976).

Although no homogeneous preparation of the mitochondrial ATPase complex have been obtained from mitochondria, it is possible to have some idea about the polypeptide composition required for reversible proton translocation (Senior, 1973; Serrano *et al.*, 1976). There are six different kinds of subunits, which are part of the F_1 moiety ($\alpha - \epsilon$ in order of decreasing molecular weight) plus the so-called F_1 inhibitor (Senior, 1973). Moreover there are at least four additional polypeptides. Two of the latter are identified as OSCP (Maclennan and Tzagoloff, 1968) and F_6 (Kanner *et al.*, 1976), both involved in binding of F_1 to the hydrophobic subunits. One of the latter is the DCCD binding protein, a proteolipid which specifically binds DCCD (Cattell *et al.*, 1971; Stekhoven *et al.*, 1972). The mitochondrial proteolipid (Criddle *et al.*, 1977) as well as the analogous polypeptide from chloroplast (Nelson *et al.*, 1977) appear to act as energy transfer inhibitor-sensitive proton holes in reconstituted systems. Proton gradients created by bacteriorhodopsin proteoliposomes were collapsed by the proteolipids, but the activity was restored by DCCD (Nelson *et al.*, 1977) or by oligomycin (Criddle *et al.*, 1977). At least one other polypeptide (approx. molecular weight : 20 000) seems to be present in F_0 preparations, but its function is not yet known.

(b) ***The complex from a thermophilic bacterium***

The purification of the mitochondrial ATPase complex has been hampered by difficulties such as the lability of purified preparations. The analogous complex from the thermophilic bacterium PS_3 appears to be much more stable (Sone *et al.*, 1975; Yoshida *et al.*, 1975). This enabled Kagawa *et al.* to make much progress with this complex. The latter has been purified to homogeneity (Sone *et al.*, 1975) and is very similar to the analogous complex of mitochondria, chloroplasts and other bacteria. Most reconstitutions have been performed using the cholate dialysis (Kagawa and Racker, 1971) or dilution technique (Racker *et al.*, 1975a,b). Upon reconsitution the purified complex displays $^{32}P_i$-ATP exchange (Sone *et al.*, 1975), ATP-dependent generation of a proton gradient (Sone *et al.*, 1976) as well as ATP-dependent membrane potential generation (Sone *et al.*, 1975, 1976, 1977). All these activities are inhibited by uncouplers and energy transfer inhibitors. The complex consists of a TF_1 moiety, which displays ATPase activity (Yoshida *et al.*, 1975), and a TF_0 moiety which appears to act as a DCCD-sensitive, TF_1-sensitive, proton-conducting pathway (Okamoto *et al.*, 1977).

TF_1 consists of five types of different subunits (α–ϵ in order of decreasing molecular weight). Due to the extraordinary stability of the enzyme, the five subunits have been isolated (Yoshida *et al.*, 1977a), and have been reassembled into functional TF_1 (Yoshida *et al.*, 1977a,b). The properties of the reconstructed TF_1 are very similar to those of the native enzyme. It possesses ATPase activity (Yoshida *et al.*, 1977a,b) and when added to TF_0 vesicles, the proton permeability of these reconstituted vesicles is reduced (Yoshida *et al.*, 1977b; Okamoto *et al.*, 1977) and the ability to catalyze $^{32}P_i$-ATP exchange is restored (Yoshida *et al.*, 1977a,b; Okamoto *et al.*, 1977). The properties of the individual subunits are quite similar to the analogous polypeptides from other sources. The study of the latter ones has of course been much more limited (Vogel and Steinhart, 1976; Nelson *et al.*, 1974; Futai *et al.*, 1974; Salton and Schor, 1972; Deters and Racker, 1975; Hockel *et al.*, 1976; Bragg and Hou, 1975; Kozlov *et al.*, 1976). In summary, δ and ϵ serve to bind the rest of TF_1 to TF_0. The γ subunit serves as a plug for the proton hole, since this subunit is able to reduce the proton permeability of TF_0 vesicles (Yoshida *et al.*, 1977b). For this to occur however, first δ and ϵ have to be bound to TF_0. The β subunit appears to be the catalytic subunit, but in order to get a minimal complex-catalysing ATP hydrolysis either $\alpha + \delta$, or γ has to be present (Yoshida *et al.*, 1977a).

In isolated TF_1 three subunits have been identified (Okamoto *et al.*, 1977). It is not clear if one of those is the DCCD-binding proteolipid. While with the mitochondrial ATPase complex it is difficult to know if upon reconstitution F_0–F_1 molecules are incorporated in the mitochondrial conformation, there is more direct evidence for TF_0–TF_1. Since the proton conductance of all functional TF_0 molecules can be blocked by TF_1, added to reconstituted TF_0 vesicles from the outside (Okamoto *et al.*, 1977), it appears that all functional TF_0 molecules are incorporated unidirectionally: namely opposite to that of the intact bacteria.

2.4.3 Reconstitution of secondary active transport systems

Two mitochondrial transport systems have been reconstituted: (a) the adenine nucleotide translocator (Shertzer and Racker, 1974, 1976) and

(b) the phosphate translocator (Shertzer *et al.*, 1977; Banerjee *et al.*, 1977). The former system catalyzes an obligatory exchange of adenine nucleotides across the inner mitochondrial membrane. Either ADP or ATP on either side of the membrane are suitable. In order to detect the exchange the reconstituted adenine nucleotide translocator vesicles are formed in the presence of unlabeled nucleotide, part of which is entrapped inside the proteoliposomes. Subsequently the external nucleotide is removed, for instance on a Dowex column (Gasko *et al.*, 1976; Shertzer and Racker, 1976). Radioactive adenine nucleotine is added and incorporation into the reconstituted vesicles can be monitored. This uptake is inhibited by well-known specific inhibitors of the system such as atractyloside and bongkrekic acid (Shertzer and Racker, 1974, 1976).

Highly purified preparations of reconstitutingly active adenine nucleotide translocator have been obtained. These preparations contain a polypeptide with a molecular weight of approximately 30 000 as a major component (Shertzer and Racker, 1976; Serrano *et al.*, 1976). From analysis of the polypeptide patterns of translocator preparations during purification, it is highly suggestive that this 30 000 molecular weight polypeptide is in fact the nucleotide translocator (Serrano *et al.*, 1976). This is consistent with the results of Riccio *et al.* who purified a homogeneous polypeptide which specifically binds [^{35}S]-carboxy-atractylate and has a molecular weight of 30 000 (Riccio *et al.*, 1975). Recently this was further confirmed. A reconstitutingly active preparation containing only the 30 000 molecular weight component has recently been obtained (Krämer and Klingenberg, 1977). When ADP and ATP are present on different sides of the membrane the translocation can be enhanced by the appropriate electrogenic conditions (Shertzer and Racker, 1976). As expected, adenine nucleotide translocation appears to be electroneutral when the same nucleotide is present on both sides of the vesicle membrane.

A first indication for the reconstitution of the phosphate transporter was obtained when partially purified adenine nucleotide translocator preparations were incorporated into lipid vesicles. Unexpectedly, with external ADP and internal ATP – thus with a negative charge moving outward – the translocation was enhanced by addition of phosphate (Shertzer and Racker, 1976). Mersalyl, a potent inhibitor of the phosphate inhibitor (Mejer *et al.*, 1970), inhibited this enhancement but not the basal rate (Shertzer and Racker, 1976). This suggested that the phosphate transporter was incorporated together with the nucleotide translocator. This was further confirmed by the following findings: purified adenine nucleotide translocator preparations did not show this phosphate stimulation upon reconsitution. Moreover, a protein fraction has been isolated (by itself not containing any nucleotide translocase activity) which, when reconsituted together with purified translocator preparations, conferred the property of mersalyl-sensitive phosphate stimulation (Shertzer *et al.*, 1977).

Using the detergent octylglucoside, a protein fraction was solubilised and partially purified, which upon incorporation into liposomes, catalyzed transport of $^{32}P_i$. $P_i \leftrightarrow P_i$ exchange were measured. Both processes were inhibited by mersalyl (Banerjee *et al.*, 1977). The following experiments demonstrate that the $P_i \leftrightarrow OH$ exchange is electrogenic. The $P_i \leftrightarrow OH$ exchange was slightly stimulated by nigericin (which can collapse the pH gradient) and by valinomycin (collapsing the membrane potential under these conditions). When both ionophores were added together, a large stimulation resulted. With phosphate also present on the inside of the proteoliposomes, $P_i \leftrightarrow P_i$ exchange occurred. The rate of this latter process was not affected by the ionophores and its rate was equal to that of the fully stimulated $P_i \leftrightarrow OH$ exchanges (Barnerjee *et al.*, 1977). This partially purified phosphate transporter was also able to confer the above described phosphate stimulation on adenine nucleotide exchange, catalyzed by purified nucleotide translocator preparations (Shertzer *et al.*, 1977).

We would like to draw the attention of the reader to a few reconstitution studies with secondary active transport proteins from other sources. Proton-coupled secondary active transport systems from bacteria have been reconstituted. Examples are the alanine carrier of the thermophilic bacterium PS_3 (Hirata *et al.*, 1976, 1977), the proline and alanine carriers from *B. subtilus* (Kusaka *et al.*, 1976) and the proline carrier of *E. coli*. Several sodium ion-coupled system have been isolated and reconstituted, such as the glucose transporter (Crane *et al.*, 1976; Kinne and Faust, 1978) and the phosphate and alanine transport systems (Kinne and Faust, 1978) from kidney. In addition, the sodium and chloride ion-coupled γ-aminobutyric acid trnasport system from brain has recently been reconsituted (Kanner, unpublished results).

2.4.4 Incorporation of multiple transport systems in the same vesicle

One of the major advantages of the reconstitution techniques is the possibility of introducing various transport systems in the same vesicle. This includes the incorporation of transport systems which are normally present in the same vesicle, such as the respiratory chain complexes and the ATPase complex of mitochondria (Racker and Kandrach, 1971, 1973; Racker *et al.*, 1975a,b; Ragan and Racket, 1973) and also incorporations of transport systems normally not residing in the same membrane, such as the co-incorporation of the mitochondrial ATPase with the light-dependent proton pump of the halophilic bacterium *H. halobium* (Racker and Hinkle, 1974). These experiments have deepened our insight into the mechanism of oxidative and photophosphorylation and provide one of the most illustrative pieces of evidence, supporting Mitchell's chemiosmotic hypothesis.

The first described example of such a 'double reconstituion' is the co-incorporation of the mitochondrial ATPase complex together with mitochondrial cytochrome oxidase (Racker and Kandrach, 1971, 1973). This represented the first report on the reconstitution of a site of oxidative phosphorylation. P/O ratios of about 0.3 were obtained. The P/O ratio of site III phosphorylation as measured in mitochondria is 1. The phosphorylation catalyzed by the proteoliposomes was found to be sensitive to uncouplers, energy transfer inhibitors and cyanide (Racker and Kandrach, 1971, 1973). These studies emphasised the importance of the asymmetric orientation of these proteins across the membrane. Under the described experimental conditions with ADP and P_i present on the outside of the reconstituted vesicles, only F_0-F_1 molecules in the submitochondrial conformation participate in the process. Site III phosphorylation was detected only when the substrate of cytochrome oxidase, cytochrome *c*, was present on the inside of the reconstituted vesicles (as in submitochondrial particles).

This asymmetry is expected on basis of the chemiosmotic hypothesis (Mitchell, 1966). According to this theory site III oxidative phosphorylation in the submitochondrial conformation may be dissected in two discrete steps. In the first step, protons are taken up concomitant with electron transfer through cytochrome oxidase. Second, the accumulated protons exit the liposomes through the ATPase complex, thereby driving ATP synthesis on the external face of the vesicle. Obviously both cytochrome oxidase as well as ATPase molecules should be in the submitochondrial conformation and present in the same lipid vesicle. As mentioned in Section 2.3.5, protons are taken up by cytochrome oxidase vesicles if cytochrome *c* is present exclusively in the internal space of the proteoliposomes. Thus functional asymmetry similar to that of submitochondrial particles has been achieved with liposomes containing both cytochrome oxidase and ATPase using a combination of internal cytochrome *c* and external ADP and phosphate. The same ATPase preparation has also been used to reconstitute site I phosphorylation using NADH dehydrogenase (Ragan and Racker, 1973), site II phosphorylation using the cytochrome b–c_1 complex (Racker *et al.*, 1975a,b) and photophosphorylation using bacteriorhodopsin from *H. halobium* (Racker and Hinkle, 1974a,b). Bacteriorhodopsin has been identified as a light-driven proton pump (Oesterhelt and Stockenius, 1973; Kanner and Racker, 1975). The latter example provides a rather suggestive example of the reconstitution technique. The ATPase complex of bovine heart is incorporated together with bacteriorhodopsin from a halophilic bacterium in vesicles made of phospholipids from soybean. It is highly improbable that, in the latter reconstiuted system, a specific direct interaction would exist between the bacteriorhodopsin and the ATPase molecules. Therefore the most likely explanation for this phenomenon of photophosphorylation is the utilisation of the proton gradient, created by bacteriorhodopsin in the light, by the ATPase to synthesise ATP.

Another example of 'double reconsitution' is the energy-dependent reversal of electron transport. When a crude transhydrogenase preparation from mitochondria was reconstituted together with the mitochondrial ATPase complex, ATP-dependent reduction of $NADP^+$ by NADH was observed (Ragan and Widger, 1975).

'Double reconstitutions' can be also utilised for secondary active transport experiments. In this case, a primary active transport system is utilised to create the appropriate ion gradient across the membrane. This ion gradient may then be utilised by the secondary active transport systems to drive solute accumulation. This concept was illustrated by Hirata *et al.* (1977). Mitochondrial cytochrome oxidase was incorporated into liposomes together with the alanine transporter of a thermophilic bacterium. Upon addition of ascorbate and cytochrome *c*, these vesicles catalyzed active alanine transport.

2.5 CONCLUDING REMARKS

Functional incorporation of membrane proteins into artificial lipid model systems has become an important tool in membrane research. The major advantages of this methodology are:

(a) Reconstitution provides a functional assay for membrane proteins. Most of these proteins promote solute transfer across biological membranes.

Therefore the functionality of the solubilised proteins can only be determined upon reconstitution.

(b) Reconstitution allows manipulation of membrane composition. This affords an opportunity to investigate membrane function and structure. Studies of this type have provided strong support for the chemiosmotic hypothesis.

(c) Incorporation of membrane proteins into lipid model systems may provide insight into the mode of membrane assembly *in vivo.*

Reconstitution is now being used in many laboratories and this approach has become an indispensable tool in membrane biochemistry.

REFERENCES

Aune, K.C., Gallagher, J.G., Gotto, A.M., Morrisett, J.D. (1977), *Biochemistry,* **16,** 2151–2156.

Awasthi, Y.C., Chuang, T.F., Keenan, T.W. and Crane, F.L. (1971), *Biochim. biophys. Acta,* **226**, 42–52.

Banerjee, R.K., Shertzer, H.G., Kanner, B.I. and Racker, E. (1977), *Biochem. biophys. Res. Commun.,* **75**, 772–778

Bangham, A.D. (1968), *Prog. Biophys. Mol. Biol.,* **18**, 29–95.

Bangham, A.D., Hill, M.W. and Miller, N.G.A., (1974), *Meth. Memb. Biol.,* **1**, 1–68.

Barenholz, Y., Gibbes, D., Litman, B.J., Goll, J., Thompson, T.E. and Carlson, F.D. (1977), *Biochemistry,* **16**, 2806–2810.

Batzri, S. and Korn, E.D. (1973), *Biochim. biophys. Acta,* **298**, 1015–1019.

Berden, J.A., Barker, R.W., Radda, G.K. (1975), *Biochim. biophys. Acta.,* **375**, 186–208.

Bergelson, L.D. and Barsukov, L.I. (1977), *Science,* **197**, 224–235.

Birchmeier, W. and Schatz, G. (1976), *Proc. natn. Acad. Sci. U.S.A.,* 4434–4338.

Boguslavsky, L.I., Kondrashin, A.A., Kozlov, V.P. and Volkov, A.G. (1975), *FEBS Letters,* **50**, 223–226.

Bragg, P.D. and Hou, C. (1975), *Arch. Biochem. Biophys.,* **167**, 311–321.

Breulet, B. and McConnell, H.M. (1976), *Biochem. biophys. Res. Commun.,* **68**, 363–368.

Brierly, C.P. and Merola, A.J. (1962), *Biochim. biophys. Acta,* **64**, 205–217.

Brunner, J., Skrabal, P. and Hauser, H. (1976), *Biochim. biophys. Acta,* **455**, 322–331.

Carroll, R. and Racker, E. (1977), *J. biol. Chem.,* **252**.

Cattell, K.J., Lindop, C.R., Knight, I.G. and Beechey, R.B. (1971), *Biochem. J.,* **125**, 169–177.

Chappell, J.B. and Crofts, A.R. (1965), *Biochem. J.,* **95**, 393–402.

Cohen, B.A. and Estabrook, R.W. (1971), *Arch. Biochem. Biophys.*, **143**, 37–65.
Colacicco, G. (1970), *Lipids*, **5**, 636–649.
Crane, R.K., Malathi, P. and Preiser, H. (1976), *FEBS Letters*, **67**, 214–216.
Criddle, R.S., Packer, L. and Shieh, P. (1977), *Proc. natn. Acad. Sci. U.S.A.*, **74**, 4306–4310.
Criddle, R.S. and Schatz, G. (1969), *Biochemistry*, **8**, 322–334.
Dawson, R.M.C., Hemington, N.L., Miller, N.G.M. and Bangham, A.D. (1976), *J. Memb. Biol.*, **29**, 179–184.
Deamer, D.W., Hill, M.W. and Bangham, A.D. (1976), *Biophys. J.*, **16**, A111.
Dean, W.L. and Tanford, C. (1977), *J. biol. Chem.*, **252**, 3551–3553.
Demel, R.A. (1974), *Meth. Enzymol.*, **32B**, 539–544.
Demel, R.A., Geurts van Kessel, W.S.M. and van Deenen, L.L.M. (1972), *Biochim. biophys. Acta*, **266**, 26–40.
Demel, R.A., Kinsky, S.C., Kinsky, C.B. and van Deenen, L.L.M. (1968), *Biochim. biophys. Acta*, **150**, 655–665.
Deters, D.W. and Racker, E. (1975), *J. biol. Chem.*, **250**, 1041–1047.
Downer, N.W., Robinson, N.C. and Capaldi, R.A. (1976), *Biochemistry*, **15**, 2930–2936.
Drachev, L.A., Jasaitis, A.A., Kaulen, A.D., Kondrashin, A.A., Chu, L.V. and Skulachev, V.P. (1976a), *J. biol. Chem.*, **251**, 7072–7076.
Drachev, L.A., Jasaitis, A.A., Kaulen, A.D., Kondrashin, A.A., Liberman, E.A., Nemecek, I.B., Ostroumov, S.A., Semenov, A. Yu. and Skulachev, V.P. (1974), *Nature*, **249**, 321–324.
Drachev, L.A., Jasaitis, A.A., Mikelsaar, H., Nemecek, I.B., Semenov, A.Y., Semenova, E.G., Severina, I.I. and Skulachev, V.P. (1976b), *J. biol.Chem.*, **251**, 7077–7082.
Enoch, H.G., Fleming, P.J. and Strittmatter, P., (1977), *J. biol. Chem.*, **252**, 5656–5660.
Eytan, G.D. and Broza, R. (1978a), *FEBS Letters*, **85**, 175–178.
Eytan, G.D. and Broza, R. (1978b), *FEBS Letters*, in press.
Eytan, G.D., Carroll, R.C., Schatz, G. and Racker, E. (1975a), *J. biol. Chem.*, **250**, 8598–8603.
Eytan, G.D., Matheson, M.J. and Racker, E., (1975b), *FEBS Letters*, **57**, 121–125.
Eytan, G.D., Matheson, M.J. and Racker, E. (1976a), *J. biol. Chem.*, **251**, 6831–6837.
Eytan, G.D. and Racker, E. (1977), *J. biol. Chem.*, **252**, 3208–3213.
Eytan, G.D., Schatz, G. and Racker, E. (1976b), In: *Structure of Biological Membranes* (Abrahamsson, S. and Pascher, I., eds), Plenum Publishing Co., New York, pp. 373–387.
Feinstein, M.B., Fernandez, S.M. and Sha'afi, R.I. (1975), *Biochim. biophys. Acta*, **413, 354–370.**
Fettiplace, R., Gordon, L.G.M., Hladky, S.B., Requena, J., Zingheim, H.P. and Haydon, D.A., (1975), *Meth. Memb. Biol.*, **4**, 1–76.
Fleischer, S., Brierly, G., Klouwen, H. and Slautterback, D.B. (1962), *J. biol. Chem.*, **237**, 3264–3272.

Futai, M., Sternweis, P.C. and Heppel, L.A. (1974), *Proc. natn. Acad. Sci. U.S.A.*, **71**, 2725–2729.

Galla, H.J. and Sackmann, E. (1975), *Biochim. biophys. Acta,* **401**, 509–529.

Gasko, O.D., Knowles, A.F., Shertzer, H.G., Suolinna, E.M. and Racker, E. (1976), *Analyt. Biochem.*, **72**, 57–65.

Goldin, S.M. (1977), *J. biol. Chem.*, **252**, 5630–5642.

Goldin, S.M. and Tong, S.W. (1974), *J. biol. Chem.*, **249**, 5907–5915.

Gollub, E.G., Trochi, P. Lin, P.K. and Sprinson, D.B. (1974), *Biochem. biophys. Res. Commun.*, **56**, 671–677.

Gregoriadis, G., Leathwood, P.D. and Ryman, B.E. (1971), *FEBS Letters,* **14**, 95–99.

Grover, A.K., Slothboom, A.J., de Haas, G.H. and Hammes, C.G. (1975), *J. biol. Chem.*, **250**, 31–38.

Guerrieri, F. and Nelson, B.D. (1975), *FEBS Letters,* **54**, 339–342.

Happe, M., Teather, R.M., Overath, P., Knobling, A. and Oesterhelt, D. (1977), *Biochim. biophys. Acta,* **465**, 411–415.

Harold, F.M. (1972), *Bact. Rev.*, **36**, 172–230.

Hartmann, W., H.-J. Galla, and Sackmann, E. (1977), *FEBS Letters,* **78**, 169–172.

Hatefi, Y., Stoggall, D.L., Galante, Y. and Hanstein, W.G. (1974), *Biochem. biophys. Res. Commun.*, **61**, 313–321.

Hauser, H., Phillips, M.C. and Stubbs, M. (1972), *Nature,* **239**, 342–344.

Hauska, G., Trebst, A. and Melandri, B.A. (1977), *FEBS Letters,* **73**, 257–262.

Hayashi, H., Vanderkooi, G. and Capaldi, R.A. (1972), *Biochem. biophys. Res. Commun.*, **49**, 92–98.

Helenius, A. and Simons, K. (1975), *Biochim. biophys. Acta,* **415**, 29–79.

Henderson, R. Capaldi, R.A. and Leigh, J.S. (1977), *J. mol. Biol.*, **112**, 631–648.

Heron, C., Corina, D. and Ragan, C.I. (1977), *FEBS Letters,* **79**, 399–403.

Hilden, S. Rhee, H.M. and Hokin, L.E. (1974), *J. biol. Chem.*, **249**, 7432–7441.

Hinkle, P.C. and Horstman, L.L. (1971), *J. biol. Chem.*, **246**, 6024–6028.

Hinkle, P.C., Kim, J.J. and Racker, E. (1972), *J. biol. Chem.*, **247**, 1338–1339.

Hirata, H., Sone, N., Yoshida, M. and Kagawa, Y. (1976), *Biochem. biophys. Res. Commun.*, **69**, 665.

Hirata, H., Sone, N., Yoshida, M. and Kagawa, Y. (1977), *J. Supramol. Struct.*, **6**, 77–84.

Hockel, M., Hulla, F.W., Risi, S. and Dose, K. (1976), *Biochim. biophys. Acta,* **429**, 1020–1028.

Hojeberg, B. and Rydström, J. (1977), *Biochem. biophys. Res. Commun.*, **78**, 1183–1190.

Holloway, P.W. and Katz, J.T. (1975), *J. biol. Chem.*, **250**, 9002–9007.

Huang, C. (1969), *Biochemistry,* **8**, 344–352.

Huang, C., Sipe, J.P., Chow, S.T. and Martin, R.B. (1974), *Proc. natn. Acad. Sci. U.S.A.*, **71**, 359–362.

Jost, P.C., Griffith, O.H., Cafaldi, R.A. and Vanderkooi, G. (1973), *Proc. natn. Acad. Sci. U.S.A.*, **70**, 480–484.

Jurtshuk, P., Sekuzu, I. and Green, D.E. (1961), *Biochem. biophys. Res. Commun.*, **6**, 76.
Kagawa, Y., Johnson, L.W. and Racker, E. (1973b), *Biochem. biophys. Res. Commun.*, **50**, 245–251.
Kagawa, Y., Kandrach, A. and Racker, E. (1973a), *J. biol. Chem.*, **248**, 676–684.
Kagawa, Y. and Racker, E. (1966), *J. biol. Chem.*, **241**, 2467–2474.
Kagawa, Y. and Racker, E. (1971), *J. biol. Chem.*, **246**, 5477–5487.
Kanner, B.I. and Racker, E. (1975), *Biochem. biophys. Res. Commun.*, **64**, 1054–1061.
Kanner, B.I., Serrano, R. Kandrach, M.A. and Racker, E. (1976), *Biochem. biophys. Res. Commun.*, **69**, 1050–1056.
Kellems, R.E. and Butow, R.A. (1972), *J. biol. Chem.*, **247**, 8043–8049.
Kinne, R. and Faust, R.G. (1978), *Biochem. biophys. Res. Commun.*, (In press).
Klappauf, K. and Schubert, D. (1977), *FEBS Letters*, **80**, 423–426.
Klingenberg, M. and Pfaff, E. (1966), In: *Regulation of Metabolic Processes in Mitochondria* (Tager, J.M. *et al.*, eds)., pp. 180–201, Elsevier, Amsterdam.
Knowles, A.F., Eytan, E. and Racker, E. (1976), *J. biol. Chem.*, **252**, 5161–5165.
Kopaczyk, K., Asai, J., Allman, D.W., Oda, T. and Green, D.E. (1968), *Arch. Biochem. Biophys.*, **123**, 602–621.
Korenbrot, J.I. (1977), *A. Rev. Physiol.*, **39**, 19–50.
Kornblatt, J.A., Chen, W.L., Hsia, J.C. and Williams, G.R. (1975), *Can J. Biochem.*, **53**, 364–370.
Kozlov, I.A., Koncrashin, A.A., Kononenko, V.A. and Metelsky, S.T. (1976), *Bioenergetics*, **8**, 1–7.
Krämer, R. and Klingenberg, M. (1977), *FEBS Letters*, **82**, 363–367.
Kremer, J.M.H., Esker, M.W.J.v.d., Pathmamoharan, C. and Wiersema, P.H. (1977), *Biochemistry*, **16**, 3932–3935.
Kuboyama, M., Yong, F.C. and King, T.E. (1972), *J. biol. Chem.*, **245**, 6375.
Kusuka, I., Hayakawa, K., Kanai, K. and Fukui, S. (1976), *Eur. J. Biochem.*, **71**, 451.
Lawford, H.G. and Garland, P.B. (1972), *Biochem. J.*, **130**, 1029–1044.
Lawford, H.G. and Garland, P.B. (1973), *Biochem. J.*, **136**, 711–720.
Lehninger, A.L. (1970), *Biochem. J.*, **119**, 129–138.
Leung, K.H. and Hinkle, P.C. (1975), *J. biol. Chem.*, **250**, 8467–8471.
London, Y., Demel, R.A., Geurts van Kessel, W.S.M., Vossenberg, F.G.A. and van Deenen, L.L.M. (1973), *Biochim. biophys. Acta*, **311**, 520–530.
London, Y., Demel, R.A., Geurts van Kessel, W.S.M., Zahler, P., and van Deenen, L.L.M. (1974), *Biochim. biophys. Acta*, **332**, 69–84.
MacLaughlin, S. and Eisenberg, M. (1975), *A. Rev. Biophys. Bioeng.*, **4**, 335–366.
MacLennan, D.H. and Tzagoloff (1968), *Biochemistry*, **7**, 1603–1610.
Mahler, H.R. (1973), *CRC Crit. Rev. Biochem.*, **1**, 381–460.
Mason, T.L., Poyton, R.O., Wharton, D.C. and Schatz, G. (1973), *J. biol. Chem.*, **248**, 1346–1354.

Matthews, F.S., Levine, M. and Argos, P. (1971), *Nature New Biol.*, **233**, 15–16.
Meijer, A.J., Groot, G.S.P. and Tager, J.M. (1970), *FEBS Letters*, **8**, 41–44.
Michaelson, D.M., Horwitz, A.F. and Klein, M.P. (1973), *Biochemistry*, **12**, 2637–2645.
Michaelson, D.M., Horwitz, A.F. and Klein, M.P. (1974), *Biochemistry*, **13**, 2605–2612.
Miller, C., Arvan, P. Telford, J.N. and Racker, E. (1976), *J. Memb. Biol.*, **30**, 276–282.
Miller, C. and Racker, E. (1976), *J. Memb. Biol.*, **26**, 319–325.
Mitchell, P. (1963), *Biochem. Soc. Symp.*, **22**, 142.
Mitchell, P. (1966), *Biol. Rev.*, **41**, 445–502.
Mitchell, P. and Moyle, J. (1967), *Biochem. J.*, **105**, 1147–1162.
Mitchell, P. and Moyle, J. (1968), *Eur. J. Biochem.*, **4**, 530–539.
Montal, M. (1974a), *Meth. Enzymol.*, **32B**, 545–554.
Montal, M. (1974b), In: *Perspectives in Membrane Biology*, pp. 591–622, Academic Press, New York.
Montal, M. (1976), *A. Rev. Biophys. Bioeng.*, **5**, 119–175.
Moyle, J. and Mitchell, P. (1973), *Biochem. J.*, **132**, 571–585.
Mueller, P., Rudin, D.O., Tien, H.T. and Wescott, W.C. (1962), *Nature*, **194**, 979–980.
Nelson, N., Eytan, E., Notsani, B.E., Sigrist, H., Sigrist-Nelson, K. and Gitler, C. (1977), *Proc. natn. Acad. Sci. U.S.A.*, **74**, 2375–2378.
Nelson, N., Kanner, B.I. and Gutnick, D.L. (1974), *Proc. natn. Acad. Sci. U.S.A.*, **71**, 2720–2724.
Oesterhelt, D. and Stockenius, W. (1973), *Proc. natn. Acad. Sci. U.S.A.*, **70**, 2853–2857.
Ohad, I. (1975), In: *Membrane Biogenesis* (Tzagoloff, A., ed.), pp. 279–350, Plenum Press, New York.
Ohnishi, S. and Ito, T. (1973), *Biochem. biophys. Res. Commun.*, **51**, 132–138.
Okamoto, H., Sone, N., Hirata, H., Yoshida, M. and Kagawa, Y. (1977), *J. biol. Chem.*, **252**, 6125–6131.
Oshino, N., Imai, Y. and Sato, R. (1971), *J. Biochem.* (Tokyo), **69**, 155–167.
Oshino, N. and Sato, R. (1971), *J. Biochem.* (Tokyo), **69**, 169–180.
Ozols, J. (1970), *J. biol. Chem.*, **245**, 4863–4874.
Ozols, J. and Gerard, C. (1977), *Proc. natn. Acad. Sci. U.S.A.*, **74**, 3725–3729.
Ozols, J., Gerard, C. and Noberga, F.G. (1976), *J. biol. Chem.*, **251**, 6767–6774.
Palmieri, F. and Klingenberg, M. (1967), *Eur. J. Biochem.*, **1**, 439–446.
Papa, S. (1976), *Biochim. biophys. Acta*, **456**, 39–84.
Papa, S., Guerrieri, F. and Lorusso, M. (1975), *Biochim. biophys., Acta*, **387**, 425–440.
Papahadjopoulos, D. and Kimelberg, H.K. (1973), In: *Progress in Surface Science* (Davison, S.G., ed.), **4**, pp. 141–232, Pergamon Press, New York.
Papahadjopoulos, D., Poste, G., Schaffer, B.E. and Vail, W.J. (1974), *Biochim. biophys. Acta*, **352**, 10–28.

Papahadjopoulos, D., Vail, W.J., Jacobson, K. and Poste, G. (1975), *Biochem. biophys., Acta,* **394**, 483–491.

Phillips, M.C. (1972), *Prog. Surf. Memb. Sci.,* **5**, 139–221.

Pressman, B.C. (1976), *A. Rev. Biochem.,* **45**, 501–530.

Racker, E. (1970), In: *Membranes of Mitochondria and Chloroplasts,* (Racker, E., ed.), pp. 127–171, Van Nostrand, New York.

Racker, E. (1972a), *J. biol. Chem.,* **247**, 8198–8200.

Racker, E. (1972b), *J. Memb. Biol.,* **10**, 221–235.

Racker, E. (1973), *Biochem. biophys. Res. Commun.,* **55**, 224–230.

Racker, E., Chien, T.F. and Kandrach, A. (1975b), *FEBS Letters,* **57**, 14–18.

Racker, E. and Hinkle, P.C. (1974), *J. Memb. Biol.,* **17**, 181–188.

Racker, E. and Kandrach, A. (1971), *J. biol. Chem.,* **246**, 7069–7071.

Racker, E. and Kandrach, A. (1973), *J. biol. Chem.,* **248**, 5841–5847.

Racker, E., Knowles, A.F. and Eytan, E. (1975a), *Ann. N.Y. Acad. Sci.,* **264**, 17–33.

Ragan, C.I. and Hinkle, P.C. (1975), *J. biol. Chem.,* **250**, 8472–8476.

Ragan, C.I. and Racker, E. (1973), *J. biol. Chem.,* **248**, 2563–2569.

Ragan, C.I. and Widger, W.R., (1975), *Biochem. biophys. Res. Commun.,* **62**, 744–749.

Reeves, J.P. and Dowben, R.M. (1970), *J. Memb. Biol.,* **3**, 123–141.

Riccio, P., Aquila, H. and Klingenberg, M. (1975), *FEBS Letters,* **56**, 133–138.

Robinson, N.C. and Capaldi, R.A. (1977), *Biochemistry,* **16**, 375–381.

Robinson, N.C. and Tanford, C. (1975), *Biochemistry,* **14**, 369–318.

Rogers, M.J. and Strittmatter, P. (1974), *J. biol. Chem.,* **249**, 5565–5569.

Rogers, M.J. and Strittmatter, P. (1975), *J. biol. Chem.,* **250**, 5713–5718.

Ross, A.H. and McConnell, H.M. (1977), *Biochem. biophys. Res. Commun.,* **74**, 1318–1325.

Rothfield, L.I. and Fried, A.A. (1975), *Meth. Memb. Biol.,* **4**, 277–292.

Rydström, J. (1977), *Biochim. biophys. Acta,* **463**, 155–184.

Rydström, J., Kanner, N. and Racker, E. (1975), *Biochem. biophys. Res. Commun.,* **67**, 831–839.

Ryrie, I.J. (1975), *Arch. Biochem. Biophys.,* **168**, 712–719.

Sadler, M.H., Hunter, D.R. and Hawarth, R.A. (1974), *Biochem. biophys. Res. Commun.,* **89**, 804–812.

Salton, M.R.J. and Schor, M.T. (1972), *Biochem. biophys. Res. Commun.,* **49**, 350–357.

Schatz, G. and Mason, T.L. (1974), *A. Rev. Biochem.,* **43**, 51–87.

Schneider, D.L., Kagawa, Y. and Racker, E. (1972), *J. biol. Chem.,* **247**, 4074–4079.

Senior, A.E. (1973), *Biochim. biophys. Acta,* **301**, 249–277.

Serrano, R., Kanner, B.I. and Racker, E. (1976), *J. biol. Chem.,* **251**, 2453–2461.

Shchipakin, V., Chuchlova, E. and Evtodienko, Y. (1976), *Biochem. biophys. Res. Commun.,* **69**, 123–127.

Sheetz, M.P. and Chan, S.I. (1972), *Biochemistry,* **11**, 4573–4581.
Shertzer, H.G., Kanner, B.I., Banerjee, R.K. and Racker, E. (1977), *Biochem. biophys. Res. Commun.,* **75**, 779–784.
Shertzer, H.G. and Racker, E. (1974), *J. biol. Chem.,* **249**, 1320–1321.
Shertzer, H.G. and Racker, E. (1976), *J. biol. Chem.,* **251**, 2446–2452.
Shimataka, T., Mihara, K. and Sato, R. (1972), *J. Biochem.* (Tokyo), **72**, 1163–1174.
Shimshick, E.J. and McConnell, H.M. (1973), *Biochem. biophys. Res. Commun.,* **53**, 446–451.
Singer, S.J. and Nicolson, G.L. (1972), *Science,* **175**, 720–730.
Slonimski, P.P. (1953), *La Formation des Enzymes Respiratoires chez la Levure,* Masson, Paris.
Sone, N., Yoshida, M., Hirata, H. and Kagawa, Y. (1975), *J. biol. Chem.,* **250**, 7917–7923.
Sone, N., Yoshida, M., Hirata, H. and Kagawa, Y. (1977), *Biochem J.,* **81**, 519–528.
Sone, N., Yoshida, M., Hirata, H., Okamoto, H. and Kagawa, Y. (1976), *J. Memb. Biol.,* **30**, 121–134.
Spatz, L. and Strittmatter, P. (1971), *Proc. natn. Acad. Sci. U.S.A.,* **68**, 1042–1046.
Spiker, R.C. and Levin, I.W. (1976), *Biochim. biophys. Acta,* **455**, 560–575.
Stekhoven, F.S., Waitkus, R.F. and Van Morkerk, H.T.B. (1972), *Biochemistry,* **11**, 1144–1150.
Strittmatter, P., Rogers, M.J. and Spatz, L. (1972a), *J. biol. Chem.,* **247**, 7188–7194.
Strittmatter, P., Rogers, M.J. and Spatz, L. (1972b), *J. biol. Chem.,* **248**, 793–799.
Swanljung, P. and Ernster, L. (1971), In: *Energy Transduction in Respiration and Photosynthesis* (Quagliariello, E., Papa, S. and Rossi, C.S., eds), Adriatica Editrice, Bari.
Swanljung, P., Frigeri, L., Ohlson, K. and Ernster, L. (1973), *Biochim. biophys. Acta,* **305**, 519–533.
Tanford, C. and Reynolds, J.A. (1976), *Biochim. biophys. Acta,* **457**, 133–170.
Thayer, W.S. and Hinkle, P.C. (1973), *J. biol. Chem.,* **248**, 5395–5402.
Thilo, L., Trauble, H. and Overath, P. (1977), *Biochemistry,* **16**, 1283–1290.
Tyrrell, D.A., Heath, T.D., Colley, C.M. and Ryman, B.E. (1976), *Biochim. biophys. Acta,* **457**, 259–302.
Tzagoloff, A., Byington, K.H. and MacLennan, D.H. (1968), *J. biol. Chem.,* **243**, 2405–2412.
Tzagoloff, A. and Meagher, P. (1971), *J. biol. Chem.,* **246**, 7328–7336.
Vail, W.J. and Riley, R.K. (1974), *FEBS Letters,* **40**, 269–273.
van Zoelen, E.J.J., Zwaal, R.F.A., Reuvers, F.A.M., Demel, R.A. and van Deenen, L.L.M. (1977), *Biochim. biophys. Acta,* **464**, 482–492.
Visser, L., Robinson, N.C. and Tanford, C. (1975), *Biochemistry,* **14**, 1194–1199.
Vogel, G. and Steinhart, R. (1976), *Biochemistry,* **15**, 208–216.
Warren, G.B., Honslay, M.D., Metcalfe, J.C. and Birdsall, N.J.M. (1975), *Nature,* **255**, 684–687.
Warren, G.B., Toon, P.A., Birdsall, N.J.M., Lee, A.G. and Metcalfe, J.C. (1974b), *Proc. natn. Acad. Sci. U.S.A.,* **71**, 622–626.

Warren, G.B., Toon, P.A., Birdsall, N.J.M., Lee, A.G. and Metcalfe, J.C. (1974a), *Biochemistry,* **13**, 5501–5507.

Wickner, W.T. (1977), *Biochemistry*, **16**, 254–259.

Wikström, M.K.F. (1977), *Nature,* **266**, 271–273.

Wikström, M.K.F. and Saari, H.T. (1977), *Biochim. biophys. Acta,* **462**, 347–361.

Yonetani, T. (1961), *J. biol. Chem.,* **236**, 1680–1688.

Yoshida, M., Okamoto, H., Sone, N., Hirata, H. and Kagawa, Y. (1977b), *Proc. natn. Acad. Sci. U.S.A.,* **74**, 936–940.

Yoshida, M., Sone, N., Hirata, H. and Kagawa, Y. (1975), *J. biol. Chem.,* **250**, 7910–7916.

Yoshida, M., Sone, N., Hirata, H. and Kagawa, Y. (1977a), *J. biol. Chem.,* **252**, 3480–3485.

Yu, C.A. and Yu, L. (1976), *Biochem. biophys. Res. Commun.,* **70**, 1115–1121.

Yu, C.A., Yu, L. and King, T.E. (1975), *J. biol. Chem.,* **250**, 1383.

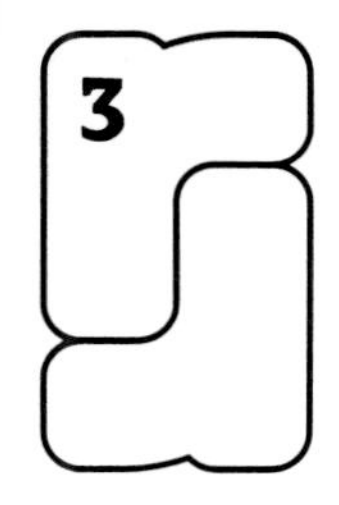

Rhodopsin: A Light-sensitive Membrane Glycoprotein

PAUL J. O'BRIEN
Laboratory of Vision Research
National Eye Institute, NIH
Bethesda, Maryland 20014 U.S.A.

Receptors and Recognition, Series A, Volume 6
Edited by P. Cuatrecasas and M.F. Greaves
Published in 1978 by Chapman and Hall, 11 New Fetter Lane, London EC4P 4EE

INTRODUCTION

Occasionally the biological world provides the inquisitive scientist with a fortuitous system that is useful for studies in many apparently unrelated disciplines. The erythrocyte is a good example. It is a cell that can be obtained in quantity and in a highly purified form. It was the obvious system in which to study oxygen binding and transport and, as such, produced a wealth of information on the chemistry of heme proteins. But it also gave us one of the first clear examples of a disease, sickle cell anemia, caused by a subtle molecular rearrangement. This observation, in turn, was evidence of an amino acid replacement resulting from a base exchange in messenger RNA; and the genetic code began to make sense. But pure hemoglobin was not the only gift of the erythrocyte. Lysis of these cells also produced a pure preparation of cell membranes. Over the years remarkable progress has been made in understanding the structure of these membranes and, by inference, all cell membranes. In short, pages could be written on the utility of red blood cells in biological research.

The vertebrate photoreceptor is another example of a cell that can provide research opportunities that reach far beyond the relatively narrow limits of the first round of studies. Those first experiments were anything but trivial and, in fact, were so significant that they earned a Nobel Prize for George Wald. He undertook the ambitious task of measuring the photochemical events that followed upon illumination of visual pigments, ultimately leading to the generation of an electrical signal. The photochemistry is now fairly well understood but our knowledge of the chemical steps in the transduction mechanism is still fragmentary. Nevertheless, photoreceptor outer segments may be the most convenient preparation of excitable membranes available, providing a nearly ideal study system. The receptor protein, rhodopsin, exhibits major spectral changes on illumination so that the primary event can be quantified. Furthermore, outer segments act as osmometers which respond to the changes in membrane permeability brought about by illumination.

But, beyond the realm of vision research, these cells may serve as model systems for the study of the role of phospholipids in membrane permeability, particularly in excitable membranes. They can also be used as the source of relatively large quantities of plasma membrane in discreet unbroken vesicles. The major protein of these membranes, rhodopsin, is a lipoprotein, a glycoprotein and a conjugated protein with a chromophore that absorbs in the visible portion of the spectrum. It is, above all, an integral membrane protein that is probably also a transmembrane protein and possibly even an ion pore under illuminating conditions.

The photoreceptor cell is remakable in so many ways that it will probably be useful to generations of investigators from dozens of disciplines. In this chapter, I will attempt to point out some of the more interesting aspects of photoreceptors and their visual pigments. The most extensive studies have been concerned with vertebrate rhodopsin and this chapter will be confined to that subject. However, another photopigment that has received considerable attention is bacteriorhodopsin, which is found in the so-called purple membrane of *Halobacterium halobium* (Henderson, 1977). Invertebrate visual pigments represent still another interesting though less intensely studied system (Abrahamson and Fager, 1973).

Although there will be some exceptions, in general no attempt will be made to review the subject exhaustively. Rather, key references will be given to cite pivotal observations or more extensive review articles on subjects not within the competence of the writer or the scope of this chapter.

3.1 ORGANIZATION OF THE VERTEBRATE PHOTORECEPTOR

The retina of a living subject can be viewed non-invasively through the use of an ophthalmoscope. Because some early neuropathologic changes can be seen in the retina, it has been called a window of the brain. And it is, in fact, an extension of the brain. The cells making up the major layers of the retina, namely the bipolar and ganglion cells, are typical neuronal elements. There are also accessory cells, the amacrine and horizontal cells, that serve to interconnect groups of cells laterally and mediate some signal processing. In addition there are glial cells which, in the retina, are referred to as Müller cells. But the truly unique cells of the retina are the photoreceptors which represent an extreme of specialization and compartmentalization. Structure is the clue to function in these cells in which the division of labor is obvious and almost absolute.

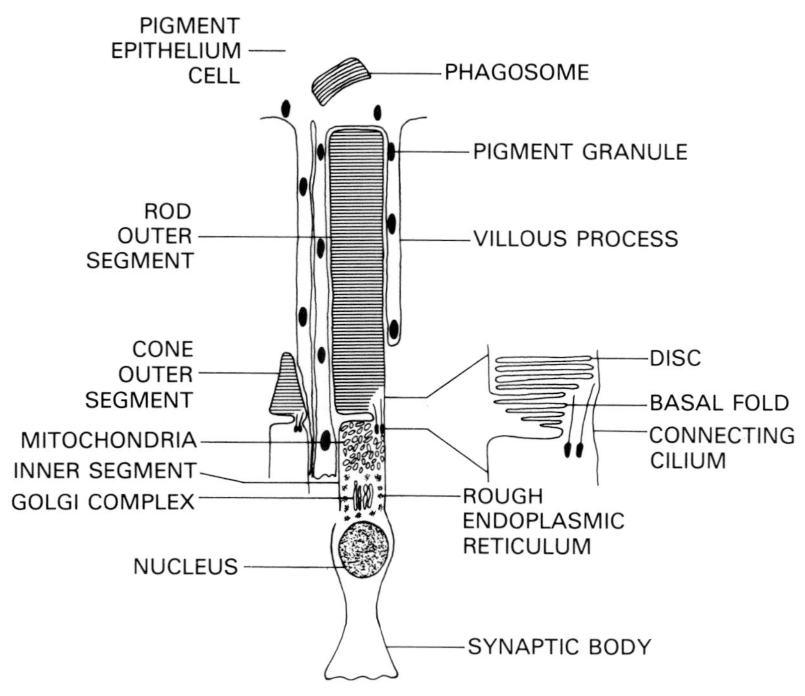

Fig. 3.1 Structure of vertebrate photoreceptors. The outer segment consists of densely packed membranous folds (cones) or discs (rods) that are derived from folds of the plasma membrane. Packets of membranes shed from the apex of the outer segment are engulfed by the adjacent pigment epithelium cells and the resulting phagosome is digested. A narrow connecting cilium joins the outer segment to the inner segment which contains the usual metabolic organelles but in a highly organized array. Light must pass through several layers of neurons as well as the inner segment before reaching the visual pigments in the membranes of the outer segments.

The peculiar architecture of photoreceptors is illustrated in Fig. 3.1. The most prominent feature of these cells is the outer segment which consists of a lamellar array of membranes containing the visual pigments. The outer segment is joined to the cell body by a narrow connecting cilium which is supported by nine double microtubules arranged circumferentially (Sjostrand, 1953). The inner segment consists of a mass of tightly packed mitochondria, which fill the space distal to the connecting cilium, and a ribosome-rich zone with a prominent Golgi complex lying between the mitochondria and the nucleus. Distal to the nucleus is a synaptic body which makes contact with the bipolar and horizontal cells. The apex of the outer segment is in intimate contact with the pigment epithelium, a single layer of cells anchored to a dense capillary bed called the choroid. Microvilli extend from the pigment epithelial cells to surround the photoreceptor outer segments. In many animals, melanin granules migrate along these villous processes under the control of microtubules and actin filaments, in response to the lighting conditions (Burnside, 1976). Incident light must first pass through the ganglion and bipolar cell layers and the synaptic, nuclear and inner segment portions of the photoreceptor before finally striking the visual pigment molecules embedded in the membranes of the outer segments. The melanin granules generally present in both the pigment epithelium and choroid then absorb excess photons and reduce back-scattering.

The distribution of metabolic activities in the photoreceptor is remarkable. In a series of elegant microdissection experiments, Lowry *et al.* (1956, 1961) demonstrated that enzyme activities such as hexokinase and malic dehydrogenase were high in the mitochondrial portion of the inner segment but were very low in the rest of the cell body. Conversely phosphofructokinase activity was found concentrated in the axonal portion of the cell between the nucleus and the synaptic body. None of these activities could be found in the outer segments which appear to have evolved almost exclusively for the purpose of trapping light and transducing that energy into an electrical signal. Many more examples of the compartmentalization of metabolic functions will emerge as we examine these unusual cells. For the present it is useful to describe the structure in greater detail.

3.1.1 Structure of rods and cones

The developmental history of photoreceptor outer segments reveals the origin of the membrane system, the mode of its continual renewal in the adult and the fundamental structures produced. As the expanded portion of Fig. 3.1 shows, the outer segment membranes are derived from infoldings of the plasma membrane that occur at the base of the outer segment where the connecting cilium is located. In development, this process is seen to occur following the formation of a knob-like protrusion supported by the microtubular bundle which emerges from the inner segment (Nilsson, 1964). However this orderly ideal is not always clearly seen. Weidman and Kuwabara (1969) found many disc membranes in developing rat retinas to be continuous with smooth endoplasmic reticulum.

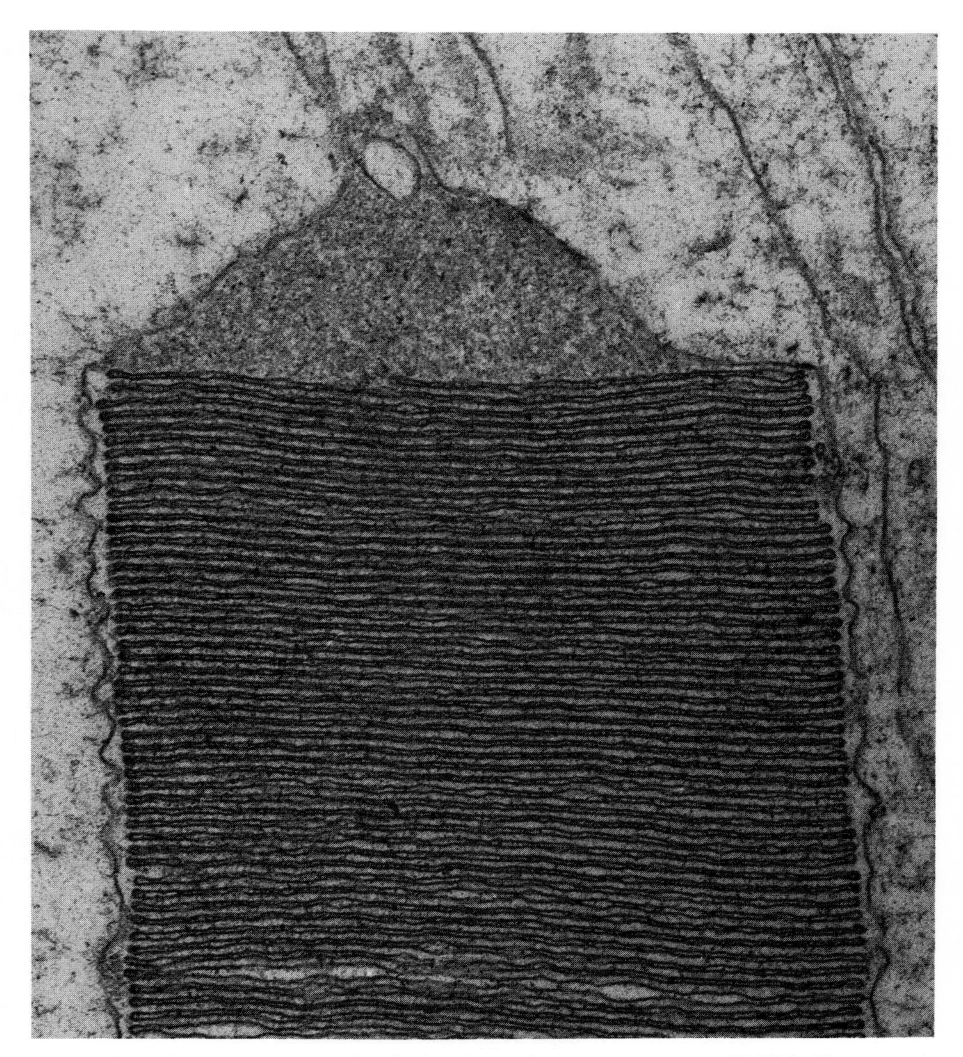

Fig. 3.2 Electron micrograph of a bovine rod outer segment x 60 000. The tip of the outer segment is surrounded by processes of the pigment epithelium. The outer segment consists of a stack of discs, flattened vesicles formed by infoldings of the plasma membrane. Courtesy of Dr Arnold I. Goldman.

Only rarely did they observe direct infoldings of the plasma membrane. The disarray of the rudimentary outer segments of these early stages of development makes ultrastructural study extremely difficult. Only after elongation of the outer segments are they packed neatly enough to permit uniform orientation of large numbers of cells.

Electron micrographs of adult Rhesus monkey retinas clearly show the basal folds of rod outer segments depicted in Fig. 3.1 (Young, 1971). They also reveal the fundamental distinction between rods and cones. At the structural level these two cell types can be distinguished on the basis of the plasma membrane-folding process. In cones, the folds remain open to the external matrix surrounding the cell. In rods, the folds ultimately are pinched off completely, forming isolated discs (Fig. 3.2), so that there is no longer a channel to the outside. Some cones show extensive pinching but always at least a small channel remains. This contact has been demonstrated in different ways. In experiments using fixed frog retinas, Cohen (1968) showed precipitates of lanthanum salts in cone outer segment folds and in the basal folds of rods but not in the remainder of the rod outer segments. Laties and Liebman (1970) used the fluorescent dye Procion Yellow, which binds to membrane proteins, to make the same point. They injected the dye into the eyes of *Necturus,* the mud puppy, and found that cone outer segments were fluorescent but rod outer segments were not, except at the base and only weakly around the plasma membrane. Since the dye could not cross intact membranes, the rods were virtually unstained whereas the cones gave free access to the dye, which bound to all the membranes so that the entire outer segment was stained. The discs of the rod outer segment were thus found to be isolated from the extracellular space. The failure of rod outer segment disc membranes to bind the dye was found not to be the result of a chemical difference in the disc membranes of rods and cones. Yoshikami *et al.* (1974) isolated frog outer segments and examined the fluorescence of didansylcystine, a dye which, like Procion Yellow, binds strongly to membranes but cannot diffuse through them. They found the anticipated results: the cone outer segments were fluorescent and intact, rod outer segments were not. However, leaky rod outer segments did take up the dye and showed intense fluorescence. The intact outer segments could be distinguished from the leaky ones by their ability to shrink in response to hypertonic conditions. All the outer segments could be made leaky by exposure to hypotonic solutions whereupon they all stained intensely.

Thus, it is firmly established that cone outer segments are essentially folded plasma membranes whereas rod outer segments have isolated each fold into a free-floating disc, which is essentially a flattened vesicle (Fig. 3.2) having a cytoplasmic surface and an interior surface, formerly the extracellular surface of the plasma membrane. An isolated rod disc is therefore an inside-out plasma membrane vesicle. This structural difference is reflected in a profound functional difference. Rods provide night vision using dim light, while cones are responsible for color vision which requires light of much greater intensity. In humans, three distinct cone visual pigments with overlapping absorption spectra are found in the blue, green and red cones.

The simple scheme depicted in Fig. 3.1 is of course not always that simple. The size and shape of photoreceptors varies considerably. In particular, a cone is not always cone-shaped and not always easy to distinguish from a rod. Nevertheless the basic structural-functional relationships that were obvious to Schultze in 1866 still persist and can be seen in most retinas of diurnal animals. In humans, the cones are concentrated near the center of focus, an area called the fovea, while rods predominate in the peripheral retina. This is the anatomical basis for the common annoying observation that, in weak light, objects can be seen more clearly if one doesn't look directly at them. The rods that are detecting the dim light are sparse at the fovea and more numerous away from the fovea, hence the greater sensitivity.

Although cones are responsible for most daylight visual activity, they are not overly numerous, since they are concentrated in one small area of the retina. As a result, rods predominate in most diurnal retinas. For this reason as well as the technical problems attending attempts at the isolation of cone outer segments and cone pigments, rod outer segment membranes have been studied almost to the exclusion of cones. Most of the discussion to follow will therefore deal with rod outer segment membranes and the rod pigment, rhodopsin. However, the comparative study of the two cell types is highly instructive and has been reviewed by Cohen (1972).

3.1.2 Localization of rhodopsin in rod outer segment membranes

Techniques have been developed to isolate and purify rod outer segments by flotation on sucrose solutions (Papermaster and Dreyer, 1974) based on the exceptionally low density of the membranes and the fact that outer segments can be broken at the connecting cilium (Fig. 3.1). All of the rhodopsin can be found in these membranes as an integral component that can be solubilized only by the use of detergents. Very few other proteins can be found intrinsic to these membranes when they are highly purified. Rhodopsin consitutes over 90% of the membrane protein (Papermaster *et al.*, 1976).

In the intact retina the rods are aligned in parallel and their tightly packed disc membranes are likewise in register (Fig. 3.2). This organization has made it possible to utilize such physical chemical probes as X-ray or neutron diffraction. Worthington (1974) concluded by X-ray analysis that rhodopsin was asymmetrically distributed in the lipid bilayer of the outer segment membrane, and he felt that rhodopsin was located at the inner surface of the disc. Freeze-fracture studies (Raubach *et al.* 1974; Corless *et al.*, 1976) likewise suggested an asymmetric localization of protein in the disc membrane, but at the outer or cytoplasmic surface of the membrane. The uncertainties of both methods demanded resolution by other approaches and there have been many.

Chemical labeling with membrane-impermeable amino group-specific reagents (Raubach *et al.*, 1974) revealed that about half the rhodopsin amino groups were accessible in rod outer segment discs, carefully prepared to prevent lysis and inversion of the membranes by resealing. Membrane-permeable reagents labeled all the amino groups. These results favored placement of rhodopsin at the cytoplasmic surface of the disc membrane as did neutron diffraction measurements (Saibil *et al.*, 1976).

The use of rhodopsin-specific antibodies coupled to peroxidase permitted Jan and Revel (1974) to demonstrate labeling of both sides of the disc membrane as well as both sides of the plasma membrane. Since their freeze-fracture experiments clearly indicated localization of particles in the cytoplasmic faces of both the disc membrane and the plasma membrane, they tentatively concluded that rhodopsin could be a transmembrane protein with the major fraction of the polypeptide located at the cytoplasmic surfaces. These observations, incidentally, point up the ontogenic relationship of the disc and plasma membranes: the discs were infoldings that became isolated, hence the cytoplasmic side of the plasma membrane remains the cytoplasmic surface of the disc (Fig. 3.1). Likewise, the existence of rhodopsin in the disc presupposes its presence in the plasma membrane which Jan and Revel found.

Proteolytic digestion has been carried out with discs that were shown to have particles in the cytoplasmic leaflet of the membrane by freeze-fracture techniques. Saari (1974) found that over one third of the rhodopsin polypeptide could be digested away by thermolysin, chymotrypsin or subtilisin without affecting the characteristic spectrum and, significantly, without losing the carbohydrate-containing portion of the molecule. Earlier studies (Heller, 1968; Shichi *et al.*, 1969) had established the glycoprotein nature of rhodopsin. Either the carbohydrate portion of rhodopsin was sequestered within the disc or in the membrane itself or it was simply not susceptible to proteolytic cleavage from the remainder of the polypeptide. However, van Breugel *et al.* (1975) while finding the same effect with pronase digestion experiments, included detergent-solubilized membranes as a control and found complete digestion of rhodopsin. Consequently, the carbohydrate must have been sequestered in some way. Both laboratories observed that trypsin failed to degrade disc membrane-bound rhodopsin at all. Trayhurn *et al.* (1975) made the same observation but, in addition, noted that plasma membranes of intact outer segments were rapidly digested away by trypsin. Since the external surface of the plasma membrane is equivalent to the internal surface of the disc membrane, there is a structural correlation with the decidedly different susceptibilities of the two membranes to trypsin. Furthermore, there is reason to believe that some of the rhodopsin molecule must be exposed to the external surface of the plasma membrane and that it might therefore actually be a transmembrane protein. Röhlich (1976) combined freeze-fracture with electron microscopic histochemistry to study the localization of the carbohydrate chain of rhodopsin. He found carbohydrate-specific reaction products inside the disc membranes by four different staining methods. With outer segment preparations broken by freezing in liquid nitrogen, he was able to demonstrate intra-membrane particles on the cytoplasmic face of the membrane by freeze-fracture. The same preparations were used to demonstrate ferritin-labeled concanavalin A binding to the inner surface of the disc membrane and the external surface of the plasma membrane. Concanavalin A is a plant lectin that binds to the non-reducing terminal α-mannoside residues of the carbohydrate chain of rhodopsin in a 1:1 ratio in disc membrane vesicles (Steinmann and Stryer, 1973). It can be concluded that rhodopsin is situated in the disc membrane in such a way that there is a substantial portion of the polypeptide exposed on the cytoplasmic surface with a carbohydrate-containing tail protruding into the intradiscal space. Godfrey (1973), using a variety of staining procedures with the electron microscope, found a carbohydrate-like substance within the discs, possibly the carbohydrate chain of rhodopsin. In the plasma membrane the carbohydrate is extracellular.

Thus rhodopsin appears to be an integral transmembrane protein with interesting similarities to glycophorin A, a component of human erythrocyte membranes. A consideration of glycophorin A is helpful in attempting to understand the architecture of red cell and possibly outer segment membranes. Glycophorin A is a polypeptide of 131 amino acids with 16 oligosaccharide chains, all of which are found on the N-terminal third of the molecule, located on the external surface of the red cell (Marchesi *et al.,* 1976). The carbohydrate content is greater than that found in rhodopsin but its location in the membrane is similar. A region of the polypeptide that passes through the lipid bilayer is composed of non-polar amino acids, many of which are hydrophobic. The C-terminal end of the molecule is enriched in glutamic and aspartic acid residues, is hydrophilic and thus could serve to bind cations or basic proteins. In fact, glycophorin A is probably associated with spectrin, a high molecular weight protein found on the inner surface of the erythrocyte plasma membrane. The C-terminal end of the molecule has been shown to extend into the cytoplasm. This was demonstrated with ferritin-labeled antibodies to a 16-amino acid peptide located near the C-terminal end of the molecule (Cotmore *et al.,* 1977). Both glycophorin A and rhodopsin qualify as ectoproteins (Rothman and Lenard, 1977), that is, integral membrane proteins with a hydrophilic portion of the protein, the carbohydrate chain, protruding on the extracellular surface. Both also appear to be transmembrane proteins with hydrophilic portions on the cytoplasmic side of the plasma membrane. In the case of rhodopsin the polarity is retained when the discs are pinched off.

3.2 STRUCTURE OF RHODOPSIN

Since rhodopsin is partially submerged in the phospholipid bilayers of the disc membrane, it would be expected to have strongly hydrophobic regions in its polypeptide structure. Heller (1968) reported that half the amino acids of rhodopsin were hydrophobic, accounting for the fact that this protein cannot be solubilized without the use of detergents. Klip *et al.* (1976) examined the distribution of these amino acids in disc membranes with a radioactive hydrophobic azide that was able to enter the non-polar interior of the membrane before being photoactivated. Upon illumination of washed membranes, the azide reacted with the hydrophobic amino acids of rhodopsin as well as with the phospholipids. Proteolytic digestion of the discs removed unlabeled peptide from the external surface and left most of the label with spectrally intact carbohydrate containing fragments. Thus, a substantial portion of the rhodopsin appears to interact with the lipid bilayer.

Molecular weight determinations have varied considerably because of incomplete purification, limitations in methodology or the anomalous behavior produced by detergent binding and by the carbohydrate chain, especially when using methods such as gel electrophoresis (Frank and

Rodbard, 1975). However the values cluster around 35 000 which is also the figure obtained by ultracentrifugal analysis of delipidated, detergent-free rhodopsin in 2-chloroethanol, a method substantially free of these drawbacks (Lewis *et al.,* 1974). Rhodopsin from quite different sources such as frog, rat and cattle all seem to have essentially the same amino acid composition (Heller, 1969). They also have in common the chromophore, 11-*cis* retinal, which is the aldehyde form of Vitamin A. In lower vertebrates, 3-dehydro-retinal can replace retinal as the chromophore (Bridges, 1972). The retinal is responsible for the characteristic absorption spectrum of rhodopsin, with a λ_{max} of about 500 nm. It is the primary light absorbing structure that initiates that photochemical events ultimately leading to an electrical signal to the brain. The structure of the chromophore and the spectral shifts induced by its interaction with the polypeptide have been extensively reviewed by Abrahamson and Wiesenfeld (1972) and by Honig and Ebrey (1974). Retinal is bound to opsin (the apoprotein of rhodopsin) through a Schiff base linkage with the ϵ-amino group of a lysine residue (Bownds, 1967). The chromophore appears to be embedded in the lipid environment of the membranes. The proteolytic studies described above all produced smaller membrane-associated polypeptides that still retained the characteristic absorption spectrum of rhodopsin, implying that the chromophore was not located at the external surface of the disc membrane. Energy transfer studies using a fluorescent probe bound to the carbohydrate of rhodopsin in the disc membranes showed that retinal was located far from the carbohydrate (Renthal *et al.,* 1973). Consequently, the chromophore cannot be at the internal surface of the disc membrane and must therefore be located in the lipid bilayer. Earlier energy transfer studies (Wu and Stryer, 1972) had suggested that rhodopsin was an elongated molecule capable of spanning the membrane.

The carbohydrate component of bovine rhodopsin was first isolated by Heller and Lawrence (1970) and later investigated in greater detail by Hargrave (1977) who found two sites of attachment of oligosaccharides. Both sites were asparagine residues located in the amino-terminal 16-amino acid peptide. The amino-terminal amino acid, methionine, was blocked and the peptide was quite hydrophilic. The location of the carbohydrate in the amino-terminal end of the molecule where protein synthesis is initiated suggests that the mechanism of synthesis and insertion of rhodopsin into the membrane is similar to that thought to occur with other ectoproteins (Rothman and Lenard, 1977) based on the 'signal hypothesis' (Blobel and Dobberstein, 1975). According to this theory, for which ample evidence is accumulating, the amino-terminal peptide contains a unique sequence of amino acids which is specifically recognized and bound to structural proteins found in the membranes of the endoplasmic reticulum. This association with the membrane proteins causes them to aggregate and form a channel in the membrane through which the peptide can pass to the lumen. The ribosome on which this peptide is elaborated becomes firmly bound to the cytoplasmic surface of the endoplasmic reticulum and polypeptide synthesis continues. Within the lumen, the nascent peptide is glycosylated. Once this occurs, the polypeptide cannot return through the membrane because of the hydrophilic carbohydrate chain. Termination of the protein with a hydrophilic peptide sequence likewise prevents it from continuing to pass completely through the membrane and it remains embedded in the membrane with a fixed polarity. Fusion of endoplasmic reticular vesicles with the plasma membrane (Palade, 1975) would cause the carbohydrate-bearing amino-terminal sequence of the protein to be extracellular, while the carboxyl terminal sequence would remain intracellular. This polarity is found in many plasma membrane proteins such as glycophorin A (Marchesi *et al.*, 1976) and vesicular stomatitis virus coat protein which is derived from the plasma membrane of infected cells by exocytosis (Lenard and Compans, 1974). The signal hypothesis and mechanism has been demonstrated elegantly for the virus coat protein (Rothman and Lodish, 1977). It is probable that this mechanism also accounts for the orientation of rhodopsin in the rod outer segment membranes.

The two carbohydrate chains isolated by Hargrave (1977) each contained two *N*-acetylglucosamine (GlcNac) residues and three mannose (Man) residues with a minor fraction (about 20%) containing five mannose residues (Fukuda *et al.*, 1977). These authors also found the basic structures to be

$$(\text{Man})_{2 \text{ or } 4}\ \alpha \rightarrow \text{Man}\ \beta \rightarrow \text{GlcNAc}\ \beta \rightarrow \text{GlcNAc}\ \beta \rightarrow \text{Asparagine}$$

These structures are commonly found in many types of glycoproteins (Kornfeld and Kornfeld, 1976) and are referred to as core oligosaccharides because they are the core upon which more complex oligosaccharides are elaborated. These two oligosaccharide chains account for most, but not all of the nine mannose and five glucosamine residues found by Plantner and Kean (1976) in purified bovine rhodopsin. The possibility remains that a third oligosaccharide chain could exist.

3.3 PROPERTIES OF ROD OUTER SEGMENT MEMBRANES

3.3.1 Lipid composition and fluidity

On a dry weight basis, the membranes of the rod outer segment are about half protein and half lipid, most of which is phospholipid. Small amounts of cholesterol (Borggreven *et al.*, 1970) and gangliosides (Hess *et al.*, 1976) have been found but the phospholipids predominate. In bovine rod outer segments, phosphatidyl choline and phosphatidyl ethanolamine each account for about 40% of the phospholipid, with phosphatidyl serine accounting for perhaps 15% (Anderson and Maude, 1970; Anderson *et al.*, 1975). Small quantities of sphingomyelin and phosphatidyl inositol make up the remainder. Remarkably similar distributions are found in the outer segments of dog, pig, human, sheep, rabbit (Anderson, 1970) and frog (Anderson and Risk, 1974). The fatty acid composition of these phospholipids is characterized by high levels of polyunsaturated fatty acids: about 30% in phosphatidyl choline, 50% in phosphatidyl ethanolamine and as much as 70% in phosphatidyl serine (Anderson *et al.*, 1975). Docosahexaenoic acid with 22 carbons and 6 double bonds is the predominant species, and it is found mainly in position 2 of the glycerol backbone of all the phospholipids (Anderson and Sperling, 1971). Stearic and palmitic acids primarily occupy position 1 except for phosphatidyl serine in which palmitic acid is replaced by another unsaturate having 24 carbons and 4 double bonds. In phosphatidyl choline, palmitic acid is also found in position 2, making this phospholipid the most highly saturated class in the membranes. Although phosphatidyl inositol is a minor component, its fatty acid composition may be indicative of an important function. In addition to palmitic, stearic and docosahexaenoic acids, it alone among the phospholipid classes has up to 18% arachidonic acid, a 20 carbon molecule with 4 double bonds.

The high level of polyunsaturated fatty acids probably insures a fluid membrane at all times, but makes the membranes particularly vulnerable to oxidative processes. However, it has been found (Dilley and McConnell, 1970; Farnsworth and Dratz, 1976) that bovine rod outer segments possess unusually high concentrations of α-tocopherol (vitamin E), an anti-oxidant. In addition, there is a high level of superoxide dismutase (Hall and Hall, 1975). Both mechanism would serve to prevent lipid peroxidation which testifies to the importance of the unsaturated fatty acids in membrane structure or function. In this regard, it may be significant that the most highly unsaturated phospholipids appear to be on the cytoplasmic surface of the disc membrane and therefore also of the plasma membrane (DeGrip, *et al.*, 1973; Litman, 1974). Thus they are able to utilize the anti-oxidant mechanisms. Molecular asymmetry is not unusual in biological membranes (Bergelson and Barsukov, 1977) and is amply demonstrated in erythrocyte membranes, where phosphatidyl choline and sphingomeyelin are found mainly in the extracellular surface while phosphatidyl ethanolamine and phosphatidyl serine are almost exclusively in the intracellular surface of the bilayer. The functional role of this arrangement can only be surmised but it seems that the more rigid, saturated phospholipids are extracellular and the more fluid, unsaturated ones are intracellular in the plasma membrane. As far as photoreceptor membranes are concerned, the fluid phospholipids occupy the same surface as the major portion of the rhodopsin molecule.

Rhodopsin has been shown to possess both mobility and rigidity in the membranes. The experimental basis for these observations rests in the linear dichroism of rod outer segments. When illuminating from the side, light that is polarized perpendicular to the axis of a single rod is absorbed approximately four times more effectively by the chromophore of rhodopsin than light polarized parallel to the axis (Harosi and MacNichol, 1974). This has been interpreted to mean that the chromophore is oriented parallel to the plane of the disc membrane which in turn is perpendicular to the axis of the rod. Thus rhodopsin does not tumble in the disc membrane. However, this dichroism can be measured only when the rod is illuminated from the side. Rods illuminated end-on show no linear dichroism presumably because of random rotational movement of the rhodopsin molecules. Brown (1972) showed that partial bleaching with polarized light failed to induce dichroism. Chromophores oriented in the plane of the polarized light would be bleached, that is, isomerized and ultimately hydrolyzed free from the opsin molecule. Chromophores oriented in the perpendicular plane would not have been bleached. Were the chromophores rigidly fixed in the membrane a significant dichroism would have been generated. Since no such effect was observed Brown assumed that the surviving rhodopsin molecules were free to rotate about an axis perpendicular to the plane of the disc membrane. He proved this by repeating the experiment with glutaraldehyde-fixed rods. In this experiment the glutaraldehyde cross-linked the membrane components so that they were no longer free to rotate. As a result, the polarized light preferentially bleached part of the rhodopsin but not that portion which was oriented perpendicular to the bleaching light. Thus rhodopsin is free to rotate in the native state and this rotation was measured directly by

Cone (1972). Using a flash photometer with a time resolution of less than 100 ns Cone showed that a transient dichroism could be measured following a bleach with a flash of polarized light. The relaxation time calculated by Cone was 20 ns and he concluded that the membrane must be highly fluid. This conclusion has been verified by a variety of experiments in which the electron spin resonance of the labeled protein (Delmelle and Pontus, 1974; Baroin *et al.*, 1977) or the nuclear magnetic resonance spectrum of the phospholipids was measured (Brown *et al.*, 1977a,b).

The highly fluid photoreceptor membrane has also been found to permit rapid lateral diffusion of rhodopsin within a single disc (Poo and Cone, 1973, 1974; Liebman and Entine, 1974). With large rod outer segments such as those in the frog or in *Necturus,* it was possible to illuminate a portion of the outer segment with a rectangular bleaching beam oriented parallel to the long axis of the rod. A measuring beam was oriented parallel to the bleaching beam so that the absorbance of the remainder of the rod could be measured. In this way, the rhodopsin in approximately half of each disc in the rod was bleached and the absorbance in the other half was measured. Within 30–45 s after bleaching, the unbleached rhodopsin had mixed completely with the bleached opsin so that both halves of the rod had the same absorption. This rapid lateral diffusion was prevented with glutaraldehyde fixation which immobilized the proteins so that one half of the rod remained bleached and the other half unbleached. When similar experiments were done with the light beams oriented perpendicular to the axis of the rod, some entire discs were bleached and some were not. No mixing occurred because the discs are separate and diffusion of rhodopsin from one to the other is prevented. Rhodopsin is therefore free to rotate and to diffuse laterally, but only within a given disc membrane and with its chromophore held parallel to the plane of the disc. This orientation optimizes the absorption of light as it passes axially through the outer segment.

3.3.2 Transduction

The unique function of rod outer segment membranes is the absorption of photons and the transduction of light energy into electrochemical energy. Photons are absorbed by the retinal chromophore of rhodopsin, triggering a chain of events that results in a transient reduction in the permeability of the plasma membrane to sodium ions. The cell is thus hyperpolarized and the release of transmitters by the synaptic body is reduced, thereby modulating the signal that is constantly produced in the dark. The details of this mechanism have proved to be elusive.

The photoreceptor inner segment is the site of a dense concentration of mitochondria (Fig. 3.1) which supply the energy for a highly active sodium pump. Sodium ions are actively pumped out of the inner segment and passively return through the plasma membrane of the outer segment and back down the connecting cilium (Hagins, 1972). This process is unimpeded in the dark, resulting in a steady level of transmitter release by the synaptic body. Illumination and bleaching of only 1 out of 10^7 rhodopsin molecules reduces the dark current of sodium ions by as much as 3% (Penn and Hagins, 1972). But that reduction can represent as many as 10^6 sodium ions (Korenbrot and Cone, 1972). In order to accomplish such an amplification, it has been postulated that an internal transmitter is released from within the discs. Rhodopsin is thought to respond to bleaching by increasing the permeability of the disc membrane to calcium ions (Hagins and Yoshikami, 1974). The calcium ions released from within the disc diffuse to the plasma membrane where they block sodium channels. The resulting reduction in the sodium ion flux into the outer segment generates a hyperpolarization because the inner segment sodium pump continues to operate. This hyperpolarization is the origin of the receptor potential which has a latency of almost 100 ms, an adequate time for the diffusion of calcium. Most of the available evidence supports this hypothesis. However the sequestering of calcium ions in the discs and their release by light has not been easy to demonstrate. Histochemically, calcium pyroantimonate precipitates within the discs have been demonstrated at the electron microscope level by Fishman *et al.* (1977). A calcium-dependent ATPase has been found in bovine rod outer segments (Sack and Harris, 1977) with the presumed function of pumping cytoplasmic calcium ions back into the discs after a bleach. The expected ATP-dependent uptake of calcium by isolated discs has also been seen (Schnetkamp *et al.*, 1977). However, although the release of some calcium by illumination of discs has been detected (Liebman, 1974), the quantities released were only a fraction of a percent of the levels demanded by the internal transmitter hypothesis (Smith *et al.*, 1977; Szuts and Cone, 1977). Perhaps technical reasons have prevented the conclusive experimental proof of the calcium theory. Nonetheless it remains a logical choice, particularly since light has been shown to increase the permeability of rhodopsin-containing artificial membranes to the extent usually produced by molecules that form transmembrane channels (Montal *et al.*, 1977).

The first descriptions of light-induced chemical changes in rhodopsin were the classic studies of Wald (reviewed in 1968) who observed a series of spectrally distinct intermediates in the bleaching process. The chromophore of rhodopsin is the 11-*cis* isomer of retinaldehyde, bound to the ε-amino group of a lysine residue. When 11-*cis* retinal absorbs a photon it is isomerized to all-*trans* retinal which is ultimately liberated from the lysine residue after passing though four or five intermediate forms. These photochemical events have been intensely studied for a clue to the transduction process. Much of this work has been reviewed by Ostroy (1977). A simplistic view of this complicated chain of events is that the first changes occur on the picosecond scale and result in the isomerization of retinal. There follow two slower spectral changes on the microsecond and millisecond scales. Finally, on a time scale of minutes, the isomerized retinal is removed from the lysine residue of opsin by hydrolysis. The events occurring on the millisecond scale involve the conversion of metarhodopsin I to metarhodopsin II where some conformational changes in the protein can be detected. At this time, the retinal-lysine linkage becomes accessible to sodium borohydride reduction. This configurational change does not involve major circular dichroism perturbations and consequently does not appear to involve alterations in the α-helix content (Shichi *et al.*, 1969). However, a birefringence transient generated in outer segment suspensions within this time frame has been attributed to the disorientation of a single phospholipid molecule for each rhodopsin bleached (Liebman *et al.*, 1974).

The relationship of these spectral intermediates to the generation of an electrochemical signal is not yet clear. But since the receptor potential is seen in a matter of milliseconds, it has been assumed that the conversion of metarhodopsin I to metarhodopsin II, having measurable conformational changes in both the protein and the phospholipid, is somehow related to the release of the internal transmitter, presumed to be calcium. It may be significant that a light-induced light scattering transient that occurs in a few milleseconds has been ascribed to a rapid shrinkage of the discs (Uhl *et al.*, 1977). However, passive efflux of calcium was ruled out since ionophores which would cause calcium to leak out of the discs could not abolish the effect of light. These authors suggested that calcium might be released from the disc membrane itself or might be released by some active process rather than through creation of a transmembrane ion pore. Furthermore, Heller *et al.* (1970, 1971) showed that disc preparations decreased in volume when illuminated. This decrease required ATP, was presumably energy-dependent, and was not caused by leaky membranes. So the release of an internal transmitter and its mode of action await more definitive demonstration. A more detailed review of transduction has been prepared by Montal and Korenbrot (1976).

3.3.3 Regeneration of rhodopsin

The all-*trans* retinal liberated during the bleaching process must be re-isomerized to 11-*cis* retinal before it can be used to regenerate a light-sensitive rhodopsin molecule. Dowling (1960) elucidated the basic steps of this recovery pathway. He found that the all-*trans* retinal was reduced to the alcohol in the photoreceptor and then transported to the pigment epithelium where it could be found mainly in the form of fatty acid esters. During dark adaptation the process was reversed and, at some point, the all-*trans* form was isomerized back to the 11-*cis* form. Whether this isomerization occurs as the ester, the alcohol or the aldehyde, and where it occurs is still not entirely clear (Bridges, 1976b, 1977). However, the regeneration of rhodopsin with 11-*cis* retinal has been effectively accomplished in artificial lipid bilayers of widely varying composition suggesting that no particular phospholipid polar head group or fatty acid chain is required, only a stable lipid bilayer (Hong and Hubbell, 1973).

3.3.4 Enzyme activities associated with rhodopsin

The presence of cyclic nucleotides in rod outer segments and the light activation of the phosphodiesterases that hydrolyze them have drawn considerable attention. Of particular interest is cyclic GMP, which is found in rod outer segments of dark-adapted frogs at the extraordinarily high level of 29 pmol/mg protein (Fletcher and Chader, 1976). Perhaps even more striking is the 10-fold decrease in cyclic GMP on light exposure of isolated outer segments or of the frogs prior to isolation of the outer segements. Cyclic AMP levels are unaffected by light and are much lower than cyclic GMP even though the reverse is generally the case. Although light causes a moderate inhibition of the synthetic enzyme, guanylate cyclase (Krishna *et al.,* 1976), the major effect of light on cyclic nucleotide metabolism is a 5–8-fold ATP-dependent increase in the activity of the phosphodiesterase which destroys the cyclic nucleotides (Miki *et al.,* 1973; Krishna *et al.,* 1976). Keirns *et al.* (1975) found that the action spectrum of phosphodiesterase activation matched the absorption spectrum of rhodopsin. They also found that 1 part of bleached disc membranes added to 99 parts of unbleached membranes could produce a half-maximal stimulation of the phosphodiesterase, and that regeneration of the bleached membranes abolished their ability to produce a stimulation. Re-bleaching restored the ability to stimulate phosphodiesterase. Sitaramayya *et al.* (1977) extracted the phosphodiesterase with Tris buffer and found that either bleached rhodopsin or fluoride ions could cause the activation. They concluded that, in the dark, the enzyme exists in an inhibited state and that bleached rhodopsin or fluoride act to remove an inhibitor in the same way that hormones and fluoride activate adenylate cyclase in many tissues. Wheeler and Bitensky (1977) have shown that GTP is the preferred nucleotide co-factor in the light activation of phosphodiesterase. It appears to be an allosteric activator rather than a phosphorylating agent and it is hydrolyzed by a GTPase which is also light-activated. The latter mechanism provides a control to restore the phosphodiesterase activity to its pre-illuminated levels so that the effects of light do not persist for long.

The role of these enzymes and of cyclic GMP in vision is not yet certain but inhibitors of phosphodiesterase accelerate the swelling of hyperosmotically shocked outer segments (Bownds and Brodie, 1975). This observation implies that an intracellular increase in cyclic nucleotides produces an increase in plasma membrane permeability, presumably to sodium ions. Brodie and Bownds (1976) were able to measure an elevation of cyclic GMP in the presence of a phosphodiesterase inhibitor, papaverine, whereas in the absence of the inhibitor they found the expected decrease in cyclic GMP after illumination, which both stimulates the phosphodiesterase and also permits blockage of the sodium channels by the internal transmitter, calcium. Lipton *et al.* (1977) found that elevated cyclic GMP levels were associated with decreased cytoplasmic calcium. They reasoned that either lowered calcium caused elevated cyclic GMP or that increases in cyclic GMP caused lowered cytoplasmic calcium, perhaps by stimulating a calcium pump. In the latter case, decreases in cyclic GMP might permit release of calcium from the discs and thereby provide a mechanism for transduction.

The intracellular decreases in cyclic nucleotide levels may or may not be rapid enough to account for the dramatic drop in sodium permeability. In either event, it is at least possible that the restoration of cyclic nucleotide levels may influence the reopening of the blocked channels.

Another enzyme activity that has attracted considerable attention is a kinase that catalyzes the phosphorylation of rhodopsin. Like the phosphodiesterase, it is active after illumination but for a different reason. The enzyme, which can be extracted from disc membranes, is not itself affected by light; rather the substrate, rhodopsin, must first be bleached before phosphorylation can occur (Frank and Buzney, 1975; Kuhn *et al.*, 1973; Miller and Paulsen, 1975; Weller *et al.*, 1975a). In experiments with bovine outer segments, there is generally about a one to one relationship between the number of rhodopsin molecules bleached and the number of phosphate molecules incorporated. However, with frog outer segments, four phosphates are incorporated for each rhodopsin bleached with bleaches in excess of 10% of the rhodopsin. At low levels of bleaching, of the order of perhaps 0.2% of the rhodopsin, up to 40 phosphate groups can be incorporated for each rhodopsin bleached (Miller *et al.*, 1977). The major difference is probably the dark adaptation state of the eyes used to prepare outer segments. Frogs can be totally dark-adapted whereas cattle eyes from a slaughterhouse are necessarily partially bleached. These results imply either multiple phosphorylation of each bleached rhodopsin or phosphorylation of many unbleached rhodopsin molecules. Incorporation of at least three phosphates in each molecule of bovine rhodopsin was suggested by the observation of Shichi *et al.* (1974), that after regeneration the phosphorylated product could be partially resolved from the bulk of the rhodopsin by chromatographic means. Miller *et al.* (1977) found that the ability to phosphorylate frog rhodopsin or opsin, as the case may be, was gradually lost if ATP was not added to outer segments immediately following partial bleaching. This decay proceeded in the dark under conditions in which no regeneration took place. Full activity was restored by a second partial bleach. It seems that some slow conformational change in the opsin molecule must take place independent of the classical regeneration process whereby a light-sensitive chromophore is added back to the opsin. This could reflect the loss of some co-operative interaction between opsin, the kinase and unbleached rhodopsin molecules. At present this question remains unresolved.

The kinase is relatively specific for opsin or rhodopsin and initial reaction rates are higher with GTP than with ATP (Chader *et al.*, 1976). Other laboratories (Miller and Paulsen, 1975; Weller *et al.*, 1970) have found ATP to be the preferred nucleotide, but since their assays were carried out for longer periods of time the light-activated GTPase (Wheeler and Bitensky, 1977) could be expected to interfere substantially. Protein kinases are generally stimulated by cyclic nucleotides but the opsin kinase is an exception (Chader *et al.*, 1976; Weller *et al.*, 1976). It is partially inhibited by cyclic AMP and is essentially unaffected by cyclic GMP which is found in high concentrations in outer segments (Fletcher and Chader, 1976).

The overall rate of phosphorylation is too slow to be involved in visual excitation and the rate of dephosphorylation (Miller and Paulsen, 1975) is even slower. Kuhn (1974) injected labeled phosphate into frogs and showed that rhodopsin obtained from these animals was unlabeled in the dark but rapidly labeled after 20 min in the light. Moreover, when the frogs were returned to the dark a slow dephosphorylation occurred with a time course resembling the dark adaptation curve, a plot of the recovery of sensitivity to dim light. It is therefore certain that phosphorylation of opsin is a physiological event and it may be that dephosphorylation is related to the restoration of light sensitivity. Conversely, phosphorylation would be expected to diminish sensitivity to light. It is known from physiological experiments (Hood and Hock, 1975) that small bleaches render photoreceptors less sensitive to further bleaches, that is, progressively greater light levels are required to produce the same electrical response on successive illuminations. This is referred to as light adaptation. The rod outer segment swelling assay was used to demonstrate the effect of inhibitors of phosphorylation on the sensitivity to light (Miller *et al.*, 1975). Light exposure prevented the swelling of outer segments in sodium chloride solutions presumably by blocking sodium channels. The extent of the suppression of swelling varied linearly with the logarithm of the light intensity (Brodie and Bownds, 1976). Light adaptation therefore occurred in this experimental system. Adenosine, an inhibitor of phosphorylation of rhodopsin, greatly enhanced the suppression of swelling by low levels of light. That is, low light levels were rendered more effective in blocking sodium channels when rhodopsin phosphorylation was prevented. Actual measurements of phosphorylation confirmed the effectiveness of the inhibitor. It can be concluded that phosphorylation of rhodopsin is related to a decrease in sensitivity to light and may be responsible for light adaptation. The mechanism of this effect could reside in the control of transmitter release. Weller *et al.* (1975b) have shown a small decrease in both the uptake and the release of calcium by outer segment discs that were illuminated in the presence of ATP. Other mechanisms are possible and it can be anticipated that many will be tested in the near future.

3.4 RENEWAL OF PHOTORECEPTOR OUTER SEGMENTS

The first indications that the outer segments were in a continuous state of turnover came from an autoradiographic experiment of Droz (1963). Since that time the specifics of this process have been elucidated in great measure because of the elegant autoradiographic studies of Young, who has reviewed his and related work (Young, 1976). At crucial points biochemical and anatomical methods of investigation have been joined together to yield information that neither approach alone could have produced.

3.4.1 Synthesis and degradation of rhodopsin

Young (1967) injected radioactive amino acids into rats, mice and frogs and visualized the labeling patterns by light microscope autoradiography. In all cases the initial labeling of the photoreceptor occurred in the inner segment. Within a day, the label had relocated to the base of the outer segment where it appeared as a band. The band of radioactivity then was progressively displaced toward the apex of the cell where it eventually disappeared. The entire process took 7–10 days in rats and mice and 5–7 weeks in frogs. The sequence of autoradiographic patterns is shown in Fig. 3.3. Young and Droz (1968) confirmed these initial observations by applying quantitative electron microscope autoradiography to frog retinas. They found that the first organelle to be labeled by the radioactive amino acids was the rough endoplasmic reticulum, the site of protein synthesis on membrane-bound ribosomes. Within an hour, the radioactivity had become concentrated in the Golgi complex where such events as glycosylation of secretory proteins and condensation and packaging of digestive enzymes occur (Palade, 1975). By two hours a band of radioactivity had begun to form at the base of the outer segment. Maximum labeling of this band was reached by four hours and it remained constant during the apical displacement of the band. In a separate study, Young (1968) showed that the labeled protein passed through the connecting cilium (Fig. 3.1) on its way to the base of the outer segment. The formation of a band of label corresponded to the formation of discrete discs from infoldings of the plasma membrane as had been seen in the early development of the outer segment discussed earlier. The maintenance of a compact band without loss of label during the apical migration implied that new discs were constantly being formed at the base of the outer segment causing the displacement of older discs. In order to maintain a constant outer segment length, it was necessary that some disposal mechanism remove the oldest discs as fast as the new ones were being made. Young and Bok (1969) followed the displacement of labeled bands to the tip of the outer segment where they became detached. They found that groups of discs were periodically pinched off and phagocytized

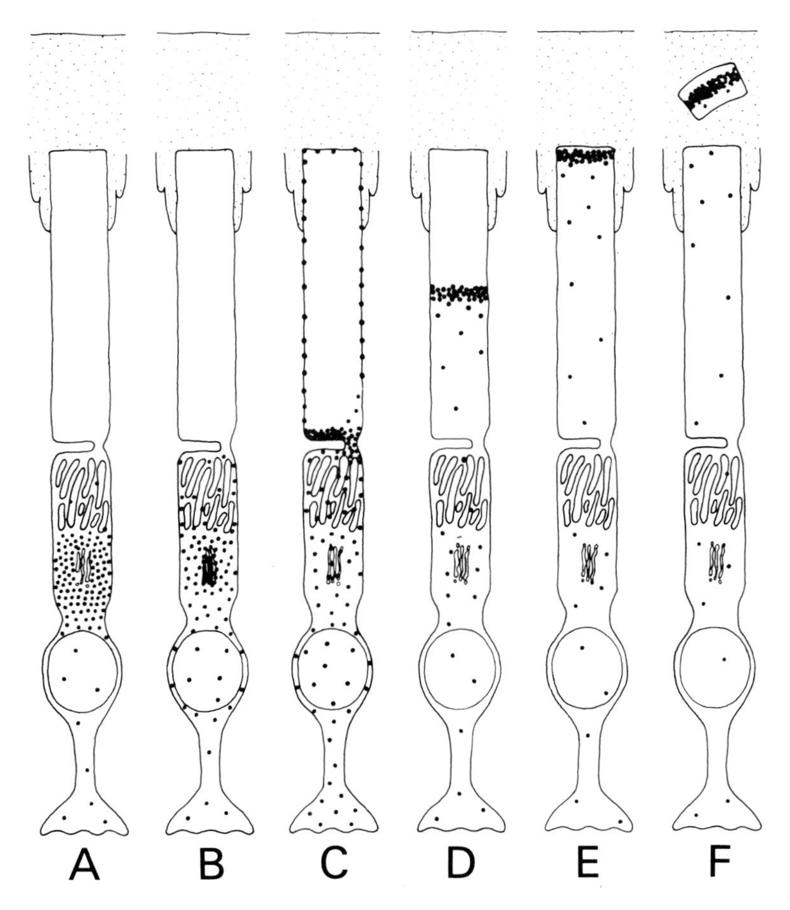

Fig. 3.3 Autoradiographic demonstration of the renewal of rod outer segments. Radioactive amino acids injected into frogs first appear in the rough endoplasmic reticulum (A) within 15 min. By 1 h (B) the newly synthesized protein is concentrated in the Golgi complex. After 2 h (C) the labeled protein has migrated to the base of the outer segment where it forms a band of radioactivity. The plasma membrane is also labeled at this time. Over the course of 5–7 weeks (D,E) the band of labeled protein is stable. New synthesis at the base of the outer segment displaces the band apically as old discs are periodically shed. Ultimately, the labeled band is shed and can be seen as a radioactive phagosome (F) which is then digested by the pigment epithelium cell (after Young, 1976).

by the adjacent pigment epithelium cells (Fig. 3.1). The labeled bands could be seen as inclusion bodies called phagosomes that were ultimately digested. The phagosomes contained many more than the 36 discs they estimated to be formed each day and therefore represented the product of several days of synthesis.

Young and Droz (1968) noticed that no band of label appeared in cone outer segments, only a diffuse, general labeling occured. Young (1969) initially thought that disc formation had stopped in cones. However, work with developing salamander cones (Young, 1975) gave similar labeling patterns when new disc synthesis was certainly taking place. In keeping with the known fluidity and continuity of these membranes, a rapid diffusion of newly synthesized protein would be expected to randomize any label rather quickly. There is, however, a transient diffuse label seen in rods in addition to the band (Fig. 3.3) (Young and Droz, 1968). It is found apical to the band, reaches a peak in about one day in frogs and declines gradually thereafter (Bok and Young, 1972). The labeling of cones follows a similar pattern, but is more intense than the weak, diffuse labeling of rods.

Although the most detailed studies have used frogs, the original observation of Young (1967) also included rats and mice. Later studies on rhesus monkey (Young, 1971) showed the phenomenon to be widespread in vertebrate retinas. La Vail (1973) has used these autoradiographic techniques to show that, during the development of mouse retina rod outer segments, the synthesis of new discs is more rapid than in the adult and the disposal of discs by shedding and phagocytosis is less than 15% of the adult level until mature rod length is reached.

It was clear that outer segment disc membranes were being labeled in all these studies but the identity of the proteins could not be determined from grains of silver. The turning point came with a crucial experiment by Hall *et al.* (1969). In a classic example of interdisciplinary science they followed the progress of band migration autoradiographically and biochemically. They purified rhodopsin from frogs injected with radioactive amino acids and simultaneously visualized the labeled band in paired animals by electron microscope autoradiography. At the time of the first appearance of radioactivity in the outer segment band, radioactivity also appeared in the purified rhodopsin. The specific activity of rhodopsin samples rose in parallel with the intensity of the outer segment band. Both reached a maximum at about four hours. The grain density of the band and the specific activity of rhodopsin remained absolutely constant for the next 8 weeks. As the labeled discs began to shed, the specific activity of rhodopsin fell. Remarkably, up to 85% of the radioactivity in the solubilized outer segments coincided with rhodopsin on the chromatographic columns used for purification. The band of labeled protein visualized autoradiographically was, in fact, rhodopsin. Moreover, the constant intensity of the band reflected the fact that, once inserted into a disc membrane, rhodopsin was completely stable and was not replaced until the entire disc was shed, phagocytized and destroyed.

The earlier stages of the life history of rhodopsin were followed by immuno-electrophoresis of extracts prepared from various subcellular fractions of retinas that were isolated at several times following injection of radioactive amino acids into frogs (Papermaster *et al.,* 1975). At all times examined, immunoprecipitable labeled opsin was associated with either the microsomal fraction or a larger membrane fragment. At no time during the migration of newly synthesized opsin to the outer segment was any found in a soluble fraction; it was always membrane-bound.

Young's work prompted efforts in my laboratory to develop an incubation system that would permit biochemical manipulation of this synthetic process. Using a Krebs–Ringer bicarbonate buffer we found that bovine retinas sustained the incorporation of labeled leucine first into microsomes and then into rod outer segments (O'Brien *et al.,* 1972). Rhodopsin was isolated and purified and found to be labeled. The labeling increased with time. We subsequently found that the major newly synthesized product in the outer segments was opsin and concluded that the addition of the chromophore, 11-*cis* retinal, was probably the final step in the synthesis of rhodopsin (O'Brien and Muellenberg, 1975). Basinger and Hall (1973), using a similar incubation system for frog retinas, were able to show leucine incorporation predominantly into outer segment rhodopsin rather than opsin. In addition, they visualized a band of label at the base of the outer segment by autoradiography. It seemed clear that retinas incubated *in vitro* were capable of carrying out at least the first steps in outer segment renewal. In later experiments with longer incubation times, Basinger *et al.* (1976) demonstrated the chemical as well as structural continuity of the outer segment plasma membrane and the basal folds. They showed autoradiographically that both were intensely labeled. They also showed chromatographically that 90% of the radioactivity was in rhodopsin and concluded that the plasma membrane as well as the basal discs contained light-sensitive visual pigment.

The subcellular site of chromphore addition was not known and the conflicting results with bovine and frog retina incubations only served to confuse. Injection of frogs with labeled retinyl acetate was followed by biochemical analyses (Hall and Bok, 1974). Light-adapted animals, with about half of their rhodopsin bleached, actually incorporated much more label into rhodopsin in short times than did dark-adapted animals. This was not surprising since regeneration of bleached rhodopsin necessarily involved transport of chromophore from the pigment epithelium (Fig. 3.1) where it was stored. Labeled chromophore in the pigment epithelium of light-adapted animals was able to become bound to opsin molecules both through new opsin synthesis and through regneration. Dark-adapted animals were restricted to new opsin synthesis. However, in these animals, no labeled chromophore appeared in rhodopsin for nearly 12 hours, even though the label was avilable in the pigment epithelium. To add to the confusion, after the first few days the label in the rhodopsin of the dark-adapted animals approached that of the light-adapted animals and both groups continued to increase in specific activity for weeks with little difference between them. Bridges and Yoshikami (1969) had observed a similar phenomenon with rats. If constant bleaching and regneration produced greater labeling at short times it should have maintained a difference at all times since regeneration of 50% bleached rhodopsin obviously involved many more molecules than those in the 36 new discs synthesized each day out of a total of perhaps 1800. The answer came from several directions. Bridges (1976a) showed that frog rod outer segments contain about a one day supply of 11-*cis* retinal for new disc synthesis. Besharse *et al.* (1977) showed that disc addition was reduced by 50% in constant darkness. These two observations thus accounted for the 12-hour lag in labeling the rhodopsin of dark-adapted animals. Then Bridges (1976b) found that in dark-adapted frog retinas there was a continual cycling of chromophore between the outer segments and the pigment epithelium. A comparable observation was made by Bok *et al.* (1977), who found that disc addition was actually a minor route of labeled chromophore incorporation into the rhodopsin of dark-adapted animals. There was an active exchange process visualized by autoradiography. One week after injection of labeled retinol the entire outer segment of dark-adapted frogs was labeled, with a significant elevation of label at the base representing new disc synthesis. This exchange process easily overcame any differences between light- and dark-adapted animals caused by bleaching and regeneration. One other significant fact became evident. At no time could any labeled chromophore be detected in inner segment structures. The retinal was added after opsin reached the outer segment and was part of the basal folds and plasma membrane. The one day supply of 11-*cis* retinal in frog outer segments also supported complete rhodopsin synthesis *in vitro* whereas in bovine retinas, which are probably more like rat retinas which have no excess stores of retinol (Dowling, 1960), opsin accumulated *in vitro*. It remains to be shown whether the pools of retinal involved in regeneration of bleached rhodopsin, synthesis of new disc membranes, and the dark exchange of chromophore are separate or common. Perhaps the existence of specific retinol receptors in retina and pigment epithelium (Futterman *et al.*, 1977; Saari and Futterman, 1976; Wiggert and Chader, 1975) may offer an experimental tool to resolve the matter. If these proteins function in retinol transport, their target membranes can be identified as was done in the case of serum retinol binding protein and the transport of retinol to pigment epithelium cells (Heller and Bok, 1976).

3.4.2 Synthesis and modification of the oligosaccharide of rhodopsin

The function of the carbohydrate portion of rhodopsin had been assumed to relate to orientation of the molecule (Heller, 1968). As discussed earlier when considering the structure of rhodopsin, this is probably quite accurate although for more complicated reasons than originally envisioned. More recently, the oligosaccharides of cell surfaces have been implicated in the fundamental behavior of cells, such as adhesion and recognition phenomena (Edelman, 1976; Roth *et al.*, 1971a,b). Since rhodopsin is a plasma membrane protein it must be assumed that it, too, could be involved in cell surface recognition phenomena, particularly in the interaction between the outer segment and the processes of the pigment epithelium cells (Fig. 3.1). A close look at the synthesis of the oligosaccharide of rhodopsin reveals similarities with major glycoprotein synthetic pathways. The composition of the oligosaccharide is simple, consisting of two kinds of sugar residues, mannose and *N*-acetylglucosamine. This is a typical core oligosaccharide and, as such, would be expected to be synthesized entirely on the rough endoplasmic reticulum as a lipid-bound oligosaccharide (Waechter and Lennarz, 1976). The mechanism of synthesis of these structures is shown in Fig. 3.4. *N*-acetylglucosamine-phosphate is first transferred to dolichol phosphate, a polyisoprenoid lipid. There then follows another transfer of *N*-acetylglucosamine from the sugar nucleotide UDP*N*-acetylglucosamine to form a lipid-linked disaccharide. Transfer of mannose from GDP mannose forms a trisaccharide. Subsequently several mannose transfers occur from GDP mannose to dolichol phosphate and then to the trisaccharide-lipid to form an oligosaccharide-lipid. The entire oligosaccharide is transfered as a unit to an asparagine residue of the polypeptide. This is the type of glycosylation reaction discussed earlier involving nascent polypeptide

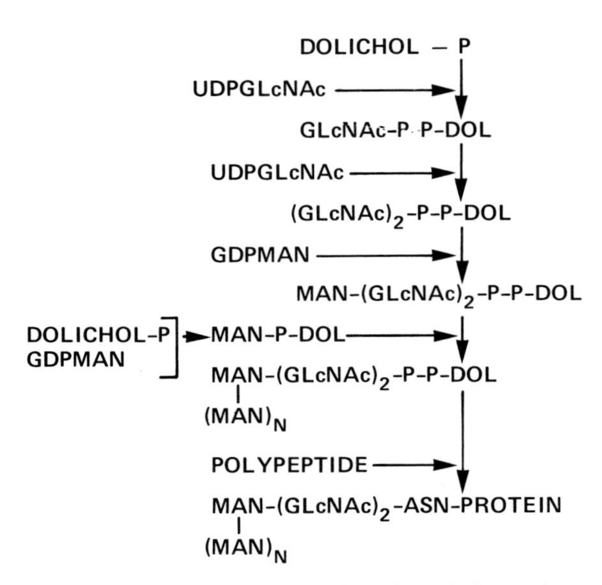

Fig. 3.4 Synthesis of lipid-linked oligosaccharide and glycosylation of newly synthesized protein at the rough endoplasmic reticulum. Sugar residues (or a sugar phosphate in the first reaction) are transferred from uridine diphosphate *N*-acetylglucosamine (UDPG1cNAc) and guanosine diphosphate mannose (GDPMan) to form a trisaccharide bound to the lipid, dolichol, through a pyrophosphate bond. Additional mannose residues are transferred from dolichol phosphate mannose (Man-P-Dol) to form the completed oligosaccharide which is transferred to the newly synthesized polypeptide to form a glycoprotein.

chains, which are being extruded into the lumen of the endoplasmic reticulum. Addition of the hydrophilic oligosaccharide prevents the polypeptide from passing back through the lipids of the membrane. There is evidence that oligosaccharide lipids are, in fact, involved in the glycosylation of rhodopsin. Kean (1977) has characterized these compounds which were synthesized by cell-free preparations of chick retina. Kean and Plantner (1976) provided evidence of a similar kind with bovine retina and found transfer of radioactive mannose to rhodopsin mediated by the lipid-linked sugars.

Both glucosamine (O'Brien and Muellenberg, 1973) and mannose (O'Brien, 1977b) can be incorporated into opsin and rhodopsin by bovine retinas incubated *in vitro*. The synthesis of core oligosaccharides by the lipid-linked pathway should be susceptible to biochemical probes such as inhibitors of protein synthesis. Since all the carbohydrate is

Table 3.1 Effect of puromycin on the incorporation of precursors of rhodopsin and opsin

Precursor	Incubation Time (h)	Specific activity, percent of control	
		Rhodopsin	Opsin
[^{3}H] Leucine	2	4.3	1.6
	4	2.5	1.1
[^{3}H] Mannose	3	9.1	3.2
	4	13	5
[^{3}H] Glucosamine	2	71	32
	4	27	12

transferred in one step to newly synthesized protein on the rough endoplasmic reticulum, a block in protein synthesis should also produce an equivalent block in the incorporation of both glucosamine and mannose into opsin. Conversely, if there are additions of sugar residues elsewhere in the cell, such as in the Golgi complex where opsin is transiently accumulated before passing to the outer segment (Young and Droz, 1968), then a block in protein synthesis might not produce a block in the incorporation of the sugars until the pool of the transiently accumulated opsin is fully glycosylated. The latter was found to be the case (O'Brien, 1977a) (Table 3.1). Puromycin, an inhibitor of protein synthesis, completely blocked the incorporation of leucine into the rhodopsin and opsin of rod outer segments. The incorporation of mannose was also effectively blocked as anticipated by the core oligosaccharide synthetic pathway (Fig. 3.4). In contrast, glucosamine incorporation was only partially blocked after two hours of incubation. But, between two and four hours there was no further incorporation. These results were consistent with the labeling of a pool of pre-existing opsin, perhaps in the Golgi complex, which was ultimately exhausted. These experiments also suggested that the carbohydrate of rhodopsin was more than a simple core oligosaccharide, which would be made entirely on the rough endoplasmic reticulum. Rather, further additions of *N*-acetylglucosamine residues took place at some time after the initial synthesis of polypeptide and core oligosaccharide. Other experiments had also suggested the same thing. In double-label incubations of bovine retinas with [^{14}C]-leucine and [^{3}H]-glucosamine, the glucosamine labeled rod outer segment rhodopsin more rapidly than did the leucine (O'Brien and Muellenberg, 1974). Thus, some glucosamine incorporation occured later in time than

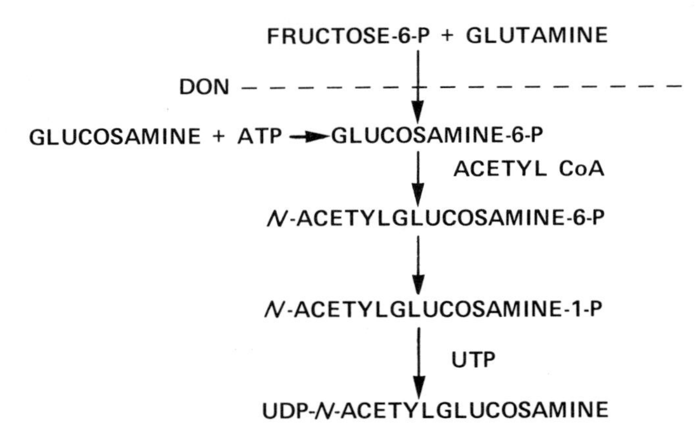

Fig. 3.5 Pathway of hexosamine metabolism and inhibition by DON. An amidotransferase converts fructose 6-phosphate to glucosamine 6-phosphate which is then acetylated with acetyl co-enzyme A. A mutase converts *N*-acetylglucosamine 6-phosphate to the 1-phosphate which is the substrate, along with uridine triphosphate (UTP), of the pyrophosphorylase that makes uridine diphosphate *N*-acetylglucosamine. An analog of glutamine, 6-diazo-5-oxo-L-norleucine (DON) blocks the first enzyme step and allows depletion of the remaining metabolites by synthetic reactions utilizing the sugar nucleotide. Exogenous radioactive glucosamine can be phosphorylated with adenosine triphosphate (ATP) by the enzyme hexokinase, bypassing the blocked enzyme step and permitting synthesis of highly labeled sugar nucleotide.

polypeptide synthesis. The timing suggested that the site was the Golgi complex.

Bok *et al.* (1974) incubated frog retinas with labeled glucosamine and found that both the endoplasmic reticulum and the Golgi complex of the photoreceptor were labeled. Clearly, there was glucosamine incorporation in the rough endoplasmic reticulum, but the Golgi labeling was not visible until 20 minutes of incubation. This might have been interpreted as the accumulation in the Golgi of protein that had been labeled on the endoplasmic reticulum. However, the labeling of that organelle was very light at the early time intervals giving rise to the suspicion that there might exist a large pool of glucosamine or its metabolic intermediates (Fig. 3.5) in the photoreceptor. This pool could have seriously diluted the labeled glucosamine so that the first molecules incorporated during the incubations might not have been of sufficiently high specific activity to be detected by autoradiography. Consequently, early labeling of the

Golgi complex could have escaped detection and, therefore, it could not be definitively stated from these experiments whether or not the Golgi complex was the site of glucosamine incorporation. This question was resolved by the use of another metabolic inhibitor, 6-diazo 5-oxo-L-norleucine (DON). This compound, an analog of L-glutamine, blocks the endogenous synthesis of glucosamine 6-phosphate (Fig. 3.5). By pre-incubating frog retinas with DON for 30 min before adding labeled glucosamine, the endogenous pools of glucosamine metabolites were used for synthetic reactions and thereby depleted because of the blocked pathway. Introduction of labeled glucosamine, which was phosphorylated by hexokinase and therefore bypassed the blocked enzyme, then resulted in the incorporation of highly labeled sugar residues, and simultaneous labeling of both the endoplasmic reticulum and the Golgi complex was visualized by autoradiography after only 10 min of incubation (Bok *et al.*, 1977). Thus, it was clear that some modification of the core oligosaccharide of opsin took place in the Golgi complex.

The Golgi apparatus is the site of addition of the so-called peripheral sugars to the core oligosaccharide of many glycoproteins, producing more complex, branched structures (O'Brien and Neufeld, 1972). Typically, three sugar residues are added sequentially to core oligosaccharides: first *N*-acetylglucosamine, then galactose and, finally, either sialic acid or fucose. It appeared that the first of these transfers had taken place. But neither galactose nor sialic acid nor fucose has ever been detected in purified rhodopsin. Nevertheless, incubation of bovine retinas with labeled galactose gave rise to labeled opsin and rhodopsin (O'Brien, 1976). The labeling pattern was similar to that found with the incorporation of glucosamine into newly synthesized opsin and rhodopsin. However, galactose incorporation was not materially affected by puromycin as glucosamine had been (O'Brien, 1977a). It thus appeared that some galactose was being incorporated into rhodopsin even later than the glucosamine incorporation that had occurred in the Golgi complex, presumably further along in the transport of opsin to the outer segment. In fact, incubation of isolated rod outer segments with labeled UDP galactose resulted in the exclusive labeling of rhodopsin with galactose (O'Brien, 1976). Similarly, fucose was transferred to rhodopsin from GDP fucose. Both sugars labeled only 1–2% of the rhodopsin molecules and thus might easily have escaped detection in carbohydrate analyses. The fact that opsin was the major labeled protein produced in the bovine retina incubations suggested that the label was entering newly synthesized opsin in the plasma membrane. When using isolated outer segments and

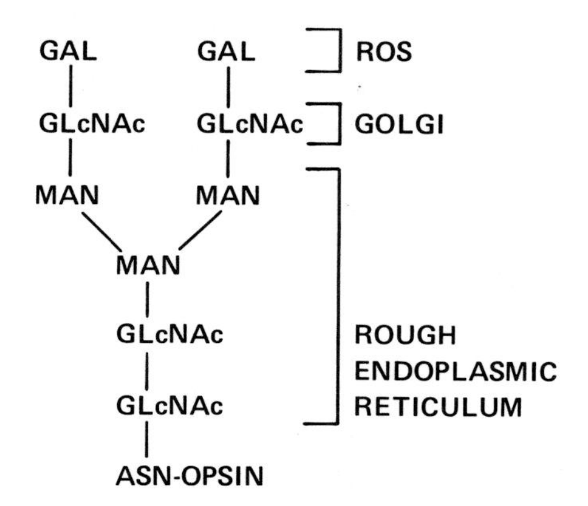

Fig. 3.6 Glycosylation of opsin and rhodopsin. A core oligosaccharide consisting of *N*-acetylglucosamine (GlcNAc) and mannose (Man) is added to opsin at the rough endoplasmic reticulum when the polypeptide is synthesized. After migration to the Golgi complex, opsin is further glycosylated with GlcNAc. It then proceeds to the outer segment where the chromophore is added along with a galactose (Gal) residue.

UDP galactose, only unbleached rhodopsin had been labeled because the freshly isolated outer segments would be expected to have chromophore in the plasma membrane rhodopsin. When new membrane was synthesized by the retina *in vitro* there was a shortage of retinal that ordinarily would have come from the pigment epithelium cells. However, these cells were not present in the incubations.

The proposed sequence of glycosylations is depected in Fig. 3.6 together with the apparent subcellular site of each reaction. It should be noted that, by analogy with other known glycoproteins (Kornfeld and Kornfeld, 1976), fucose could be bound either to the terminal galactose or to the *N*-acetylglucosamine residue linked to asparagine.

The presence of some galactose or fucose in the plasma membrane rhodopsin raises the question of a possible function (O'Brien, 1976). This function may relate to the phagocytosis of packets of discs periodically shed from the tip of the outer segment. Once these packets, which are surrounded by plasma membrane, are shed, they are engulfed by pigment epithelial cells and digested (Young and Bok, 1969). Intact rod outer segments are not attacked by pigment epithelial cells. The transient presence of a different carbohydrate structure could provide a unique surface marker enabling the pigment epithelium to distinguish between packets of shed discs and intact outer segments. The extra sugars not found in the bulk of the rhodopsin may be added in preparation for shedding. Alternatively, they could be added to all new plasma membrane rhodopsin molecules and selectively removed as the packet of discs to be shed is pinched off by an infolding of the plasma membrane near the tip of the outer segment. A similar type of infolding occurs at the base of the outer segment in the formation of new discs and could involve the removal of the extra sugars by a common mechanism. In this regard, the pinching off of synaptic vesicles from the plasma membrane, which occurs at the other end of the same cell, is known to result in the removal of carbohydrate from the membrane (Heuser and Reese, 1973), and may represent yet another use of a common mechanism. By either mechanism, the selective addition or removal of sugars would create a specific difference in the carbohydrate structures between the surfaces of the plasma membranes of the intact outer segment and of the shed packet of discs. This difference could be recognized by the pigment epithelial cells just as the presence of fucose residues on the cell surface glycoproteins of lymphocytes is recognized by reticulo-endothelial cells. Removal of fucose abolishes the selective homing of lymphocytes to reticulo-endothelial tissue where they are ordinarily removed from blood and transported to lymph ducts (Gesner and Ginsburg, 1964). A similar recognition mechanism could also account for phagocytosis of shed outer segment tips, but this hypothesis awaits definitive proof.

The transfer of galactose from UDP galactose to rhodopsin may involve an intermediate compound, further distinguishing this reaction from the typical Golgi complex galactosyl transferase reaction, which is a direct, one-step transfer. Nucleotide competition experiments suggest the existence of a high-energy intermediate such as a sugar-lipid donor (O'Brien, 1978). One possibility is retinyl phosphate galactose (Helting and Peterson, 1972). When retinyl phosphate was added to incubation mixtures containing outer segments and UDP galactose, there was a 50% stimulation in the transfer of galactose to rhodopsin (O'Brien, 1978). Direct transfer from retinyl phosphate galactose to rhodopsin has not yet been shown. However, since this glycosylation seems to take place on the extracellular surface of the plasma membrane, a lipid-soluble donor may be required to transport an activated sugar molecule from the cytoplasm to the exterior of the cell. An interesting alternative is the possibility that the soluble retinol receptor protein found in retina and pigment epithelium (Wiggert and Chader, 1975) could be a retinyl phosphate or retinyl phosphate galactose transport protein. Thus the pigment epithelium could supply the co-factor and, in fact, control the glycosylation of plasma membrane rhodopsin and perhaps even control the shedding process. The chemical events in shedding and phagocytosis could turn out to be of universal interest in understanding membrane recognition and function.

3.4.3 Synthesis and turnover of rod outer segment phospholipids

Much less work has been done on the renewal of outer segment phospholipids than on the renewal of proteins. Bibb and Young (1974a) injected labeled glycerol into frogs and followed the labeling of outer segments by autoradiography. Since glycerol is the basic structure upon which all the phospholipids are built, it reflects the synthesis of any or all phospholipids. The labeled glycerol first appeared in the inner segment, then formed a band of radioactivity at the base of the outer segment indicative of new membrane synthesis. As such it resembled the rhodopsin labeling sequence. However labeled glycerol also became distributed throughout the outer segment indicating that individual phospholipid molecules could be removed from the disc membranes and replaced. Possibly phospholipid exchange proteins could be involved in this replacement process (Wirtz, 1974). Furthermore, when fatty acid precursors of phospholipids were injected into frogs a much more rapid general labeling of the outer segment took place (Bibb and Young, 1974b). Apparently components of the phospholipids can be replaced by exchange reactions. Likewise, choline was rapidly distributed in the outer segment and must also be replaced by an exchange mechanism (Basinger and Hoffman, 1976). In this case, it was shown that the choline must first be processed by the inner segment, presumably through the formation of phosphoryl choline and CDP choline. Light caused a significant inhibition of choline incorporation but had no effect on ethanolamine incorporation. The significance of this observation is not clear, but it could imply a differential replacement of some phospholipid. Perhaps it relates to the unpublished observation of Young and Basinger that the tips of some rods became intensely labeled when frog retinas were incubated with radioactive choline. This observation suggests a chemical change in those membranes about to be shed and it may represent a dark reaction preceeding the shedding which is subsequently triggered by light (Basinger *et al.*, 1976). This phenomenon will be discussed below.

Hall *et al.* (1973) injected labeled phosphate into frogs and isolated the outer segment phospholipids. The major classes, phosphatidyl choline, phosphatidyl serine, phosphatidyl ethanolamine and sphingomyelin, were all gradually labeled over the course of three weeks. However, phosphatidyl inositol, a minor component of the outer segment, was very rapidly labeled by five days then rapidly dropped in specific activity until it resembled the other phospholipids. Presumably, this reflects an active role in the function of the membrane.

3.4.4 The effects of light on the renewal process

For almost a decade, the renewal of rod outer segments was thought to be a continual process of disc synthesis at the base of the outer segment and periodic, balanced removal of the oldest discs from the apex. However, LaVail (1976a,b) found that the shedding events were not simply random expressions of the achievement of maximal rod length but were, in fact, synchronized by the daily light cycle. In rats kept on a regular cycle of 12 hours of darkness and 12 hours of light, the number of phagosomes in the pigment epithelial cells, that is, the shed outer segment tips, increased four-fold during the first two hours after the onset of light. By the fourth hour of light most of the phagosomes had been digested and their numbers dropped to a very low level for the remainder of the light period. A few phagosomes could be found during the subsequent dark period but the next burst of shedding occurred only after the lights were turned on. Earlier illumination did not produce an earlier peak of shedding. A full 12 hours of darkness appeared to be necessary before light could induce shedding. Furthermore, a burst of shedding occurred on schedule even when the lights were not turned on, suggesting a circadian rhythm entrained by the previous light cycle. In contrast, no such ciradian rhythm was found in adult frog disc shedding (Basinger *et al.*, 1976). In this species, light seemed to be required to initiate the burst of shedding. In constant darkness, the outer segments became longer and there was a low level of random shedding. After several days of darkness, light triggered a larger burst of shedding than that seen during a normal day/night cycle. Normally frog rods shed once every four days. Apparently, the longer dark period caused a greater number of rod outer segments to become primed. The nature of the dark priming is not known nor is the mechanism by which light, or the biological clock in the case of rats, triggers the shedding. Perhaps it is related to the oligosaccharide changes discussed earlier.

A pattern of shedding similar to that of adult frogs was seen in tadpoles (Hollyfield *et al.*, 1976). In addition to the light-triggered shedding of large phagosomes, there was continuous shedding of small groups of discs that occurred in both light and darkness. In constant darkness, the quantity of rhodopsin increased with increasing rod length in both adult frogs and tadpoles (Bridges *et al.*, 1976). Moreover, when massive shedding was induced by light, the loss of rhodopsin was consistent with the shortening of outer segments. However, a rapid recovery of rhodopsin content was noted after the massive shedding. The rate of recovery was 3–4 times greater than the average daily outer segment renewal rate previously measured (Young, 1967). Subsequently, Besharse *et al.* (1977b) found that in frogs there was greater synthesis of new discs in constant light than there was in cyclic light and that there was less in constant darkness. Using tadpoles of the clawed toad, *Xenopus laevis*, Besharse *et al.* (1977a,b) found by autoradiography that the renewal rate was 5 times greater in the first 8 hours of a normal diurnal cycle than during the following 16 hours. Moreover, the number of basal folds (Fig. 3.1) increased from three to eleven during the 8 hours of rapid synthesis then decreased to three by the beginning of the dark period at 12 hours. Consequently, there was biochemical, autoradiographic and ultrastructural evidence of an accelerated period of membrane synthesis following the onset of light. Thus, both shedding and synthesis can be stimulated by light in *Xenopus*.

Besharse *et al.* (1977a) were able to affect synthesis and shedding independently by modifying the lighting conditions. In constant light, a high rate of disc addition was maintained while shedding was dramatically reduced producing a rapid elongation of outer segments. On the other hand, when animals were maintained on a daily cycle of 2.5 h light and 21.5 h darkness, disc addition was greatly reduced but shedding continued at nearly normal levels and the outer segments became shorter. The stimulation of synthesis required prolonged lighting whereas the triggering of shedding required a brief exposure to light and the two effects may not be directly related. In constant darkness, both disc addition and shedding were nearly halved for the first two days. However, by the third day a nearly normal peak of phagosomes was found. It occurred several hours later than it would have in cyclic light. A similar dark shedding had been observed in the tadpoles of frogs (Hollyfield *et al.*, 1976) and may represent a circadian rhythm temporarily delayed because of the slower synthesis of discs in constant darkness.

The factors controlling synthesis and shedding are not clearly understood. Light accelerates synthesis but is not necessary for synthesis. Light synchronizes shedding but does not seem to be totally necessary for shedding in either rats or amphibians. Darkness, on the other hand appears to be essential as a prelude for shedding. Whether the dark 'priming' events involve reactions wholly within the retina or require humoral, perhaps endocrine, factors such as melatonin and serotonin remains to be clarified. This is an area of intensive research that promises to develop rapidly. Any mechanism proposed to account for the control of shedding will have to include cone shedding which has been demonstrated in humans (Hogan *et al.*, 1974) and in five species of squirrels (Anderson and Fisher, 1976). The relationship of cone shedding to rod shedding is confused by the recent discovery that cone shedding occurs early in the dark period of the daily light cycle in lizards (Young, 1977a) chicks (Young, 1977b) and goldfish (O'Day and Young, 1977). It appears that shedding occurs during that part of the light cycle when the cell would not normally be expected to function as a photoreceptor. Thus, the rods shed in bright light when they are overwhelmed by light and are not active and the cones shed when the light levels are too low for them to function. The visual pigments are bleached in rods and unbleached in cones when shedding takes place. The resolution of these questions will be interesting.

REFERENCES

Abrahamson, E.W. and Fager, R.S. (1973), In: *Current Topics in Bioenergetics* (Sanadi, D.R., ed.), Vol. 5, p. 125, Academic Press, New York.

Abrahamson, E.W. and Wiesenfeld, J.R. (1972), In: *Handbook of Sensory Physiology. Photochemistry of Vision* (Dartnall, H.J.A., ed.), Vol. VII Part 1, p. 69, Springer-Verlag, New York.

Anderson, D.H. and Fisher, S.K. (1976), *J. Ultrastruct. Res.*, **55**, 119.

Anderson, R.E. (1970), *Exp. Eye Res.*, **10**, 339.

Anderson, R.E. and Maude, M.B. (1970), *Biochemistry*, **9**, 3624.

Anderson, R.E., Maude, M.B. and Zimmerman, W. (1975), *Vision Res.*, **15**, 1087.

Anderson, R.E. and Risk, M. (1974), *Vision Res.*, **14**, 129.

Anderson, R.E. and Sperling, L. (1971), *Arch. Biochem. Biophys.*, **144**, 673.

Baroin, A., Thomas, D.B., Osborne, B. and Devaux, P.F. (1977), *Biochem. biophys. Res. Commun.*, **78**, 442.

Basinger, S.F., Bok, D. and Hall, M. (1976), *J. Cell Biol.*, **69**, 29.

Basinger, S.F. and Hall, M.O. (1973), *Biochemistry*, **10**, 1996.

Basinger, S. and Hoffman, R. (1976), *Exp. Eye Res.*, **23**, 117.

Basinger, S., Hoffman, R. and Matthes, M. (1976), *Science*, **194**, 1074.

Bergelson, L.D. and Barsukov, L.I. (1977), *Science*, **197**, 224.

Bersharse, J.C., Hollyfield, J.G. and Rayborn, M.E. (1977a), *J. Cell Biol.*, **75**, 507.

Berharse, J.C., Hollyfield, J.G. and Rayborn, M.E. (1977b), *Science,* **196**, 536.
Bibb, C. and Young, R.W. (1974a), *J. Cell Biol.,* **62**, 378.
Bibb, C. and Young, R.W. (1974b), *J. Cell Biol.,* **61**, 327.
Blobel, G. and Dobberstein, B. (1975), *J. Cell Biol.,* **67**, 835.
Bok, D., Basinger, S.F. and Hall, M.O. (1974), *Exp. Eye Res.,* **18**, 225.
Bok, D., Hall, M.O. and O'Brien, P.J. (1977), In: *International Cell Biology 1976–1977* (Brinkley, B.R. and Porter, K.R., eds), p. 608, Rockefeller University Press, New York.
Bok, D. and Young, R.W. (1972), *Vision Res.,* **12**, 161.
Borggreven, J.M.P.M., Daemen, F.J.M. and Bonting, S.L. (1970), *Biochim biophys. Acta,* **202**, 374.
Bownds, D. (1967), *Nature,* **216**, 1178.
Bownds, D. and Brodie, A.E. (1975), *J. gen. Physiol.,* **66**, 407.
Bridges, C.D.B. (1972), In: *Handbook of Sensory Physiology. Photochemistry of Vision* (Dartnall, H.J.A., ed.), Vol. VII Part 1, p. 417, Springer-Verlag, New York.
Bridges, C.D.B. (1976a), *Nature,* **259**, 247.
Bridges, C.D.B. (1976b), *Exp. Eye Res.,* **22**, 435.
Bridges, C.D.B. (1977), *Exp. Eye Res.,* **24**, 571.
Bridges, C.D.B., Hollyfield, J.G., Besharse, J.C. and Rayborn, M.E. (1976), *Exp. Eye Res.,* **23**, 637.
Bridges, C.D.B. and Yoshikami, S. (1969), *Nature,* **221**, 275.
Brodie, A.E. and Bownds, D. (1976), *J. gen. Physiol.,* **68**, 1.
Brown, M.F., Miljanich, G.P. and Dratz, E.A. (1977a), *Proc. natn. Acad. Sci. U.S.A.,* **74**, 1978.
Brown, M.F., Miljanich, G.P. and Dratz, E.A. (1977b), *Biochemistry,* **16**, 2640.
Brown, P.K. (1972), *Nature New Biol.,* **236**, 35.
Burnside, M.B. (1976), *Exp. Eye Res.,* **23**, 257.
Chader, G.J., Fletcher, R.T., O'Brien, P.J. and Krishna, G. (1976), *Biochemistry,* **15**, 1615.
Cohen, A.I. (1968), *J. Cell Biol.,* **37**, 424.
Cohen, A.I. (1972), In: *Handbook of Sensory Physiology. Physiology of Photoreceptor Organs* (Fuortes, M.G.F., ed.), Vol. II Part 2, p. 63, Springer-Verlag, New York.
Cone, R.A. (1972), *Nature New Biol.,* **236**, 39.
Corless, J.M., Cobbs, W.H., Costello, M.J. and Robertson, J.D. (1976), *Exp. Eye Res.,* **23**, 295.
Cotmore, S.F., Furthmayer, H. and Marchesi, V.T. (1977), *J. mol. Biol.,* **113**, 539.
DeGrip, W.J., Daemen, F.J.M. and Bonting, S.L. (1973), *Biochim. biophys. Acta,* **323**, 125.
Delmelle, M. and Pontus, M. (1974), *Biochim. biophys. Acta,* **365**, 47.
Dilley, R.A. and McConnell, D.G. (1970), *J. Memb. Biol.,* **2**, 317.
Dowling, J.E. (1960), *Nature,* **188**, 114.

Droz, B. (1963), *Anat. Rec.*, **145**, 157.
Edelman, G.M. (1976), *Science*, **192**, 218.
Farnsworth, C.C. and Dratz, E.A. (1976), *Biochim. biophys. Acta*, **443**, 556.
Fishman, M.L., Oberc, M.A., Hess, H.H. and Engel, W.K. (1977), *Exp. Eye Res.*, **24**, 341.
Fletcher, R.T. and Chader, G.J. (1976), *Biochem. biophys. Res. Commun.*, **70**, 1297.
Frank, R.N. and Buzney, S.M. (1975), *Biochemistry*, **23**, 5110.
Frank, R.N. and Rodbard, D. (1975), *Arch. Biochem. Biophys.*, **171**, 1.
Fukuda, M.N., Papermaster, D.S. and Hargrave, P.A. (1977), *Fedn. Proc. fedn. Am. Socs. exp. Biol.*, **36**, 895.
Futterman, S., Saari, J.C. and Blair, S. (1977), *J. biol. Chem.*, **252**, 3267.
Gesner, B.M. and Ginsburg, V. (1964), *Proc. natn. Acad. Sci. U.S.A.*, **52**, 750.
Godfrey, A.J. (1973), *J. Ultrastruct. Res.*, **43**, 228.
Hagins, W.A. (1972), *A. Rev. Biophys. Bioeng.*, **1**, 131.
Hagins, W.A. and Yoshikami, S. (1974), *Exp. Eye Res.*, **18**, 299.
Hall, M.O., Basinger, S.F. and Bok, D. (1973), In: *Biochemistry and Physiology of Visual Pigments* (Langer, H., ed.), p. 319, Springer-Verlag, New York.
Hall, M.O. and Bok, D. (1974), *Exp. Eye Res.*, **18**, 105.
Hall, M.O., Bok, D. and Bacharach, A.D.E. (1969), *J. mol. Biol.*, **45**, 397.
Hall, M.O. and Hall, D.O. (1975), *Biochem. biophys. Res. Commun.*, **67**, 1199.
Hargrave, P.A. (1977), *Biochim. biophys. Acta*, **492**, 83.
Harosi, F.I. and MacNichol, Jr., E.F. (1974), *J. Opt. Soc. Am.*, **64**, 903.
Heller, J. (1968), *Biochemistry*, **7**, 2906.
Heller, J. (1969), *Biochemistry*, **8**, 675.
Heller, J. and Bok, D. (1976), *Am. J. Ophthal.*, **81**, 93.
Heller, J. and Lawrence, M.A. (1970), *Biochemistry*, **9**, 864.
Heller, J., Ostwald, T.J.and Bok, D. (1970), *Biochemistry*, **9**, 4884.
Heller, J., Ostwald, T.J.and Bok, D. (1971), *J. Cell Biol.*, **48**, 633.
Helting, T. and Peterson, P.A. (1972), *Biochem. biophys. Res. Commun.*, **46**, 429.
Henderson, R. (1977), *A. Rev. Biophys. Bioeng.*, **5**, 87.
Hess, H.H., Stoffyn, P. and Sprinkle, K. (1976), *J. Neurochem.*, **26**, 621.
Heuser, J.E. and Reese, T.S. (1973), *J. Cell Biol.*, **57**, 315.
Hogan, M.J., Wood, I. and Steinberg, R.H. (1974), *Nature*, **252**, 305.
Hollyfield, J.G., Besharse, J.C. and Rayborn, M.E. (1976), *Exp. Eye Res.*, **23**, 623.
Hong, K. and Hubbell, W.L. (1973), *Biochemistry*, **12**, 4517.
Honig, B. and Ebrey, T.G. (1974), *A. Rev. Biophys. Bioeng.*, **3**, 151.
Hood, D.C. and Hock, P.A. (1975), *Vision Res.*, **15**, 545.
Jan. L.Y. and Revel, J-P. (1974), *J. Cell Biol.*, **62**, 257.
Kean, E.L. (1977), *J. biol. Chem.*, **252**, 5622.
Kean, E.L. and Plantner, J.J. (1976), *Exp. Eye Res.*, **23**, 89.
Keirns, J.J., Miki, N., Bitensky, M.W. and Keirns, M. (1975), *Biochemistry*, **14**, 2760.
Klip, A., Darszon, A. and Montal, M. (1976), *Biochem. biophys. Res. Commun.*, **72**, 1350.

Korenbrot, J.I. and Cone, R.A. (1972), *J. gen. Physiol.*, **60**, 20.
Kornfeld, R. and Kornfeld, S. (1976), *A. Rev. Biochem.*, **45**, 217.
Krishna, G., Krishnan, N., Fletcher, R.T. and Chader, G. (1976), *J. Neurochem.*, **27**, 717.
Kuhn, H. (1974), *Nature,* **250**, 588.
Kuhn, H., Cook, J.H. and Dreyer, W.J. (1973), *Biochemistry,* **12**, 2495.
Laties, A.M. and Liebman, P.A. (1970), *Science,* **168**, 1475.
Lavail, M.M. (1973), *J. Cell Biol.*, **58**, 650.
LaVail, M.M. (1976a), *Exp. Eye Res.*, **23**, 277.
LaVail, M.M. (1976b), *Science,* **194**, 1071.
Lenard, J. and Compans, R.W. (1974), *Biochim. biophys. Acta,* **344**, 51.
Lewis, M.S., Krieg, L.C. and Kirk, W.D. (1974), *Exp. Eye Res.*, **18**, 29.
Liebman, P.A. (1974), *Invest. Ophthal.*, **13**, 700.
Liebman, P.A. and Entine, G. (1974), *Science,* **185**, 457.
Liebman, P.A., Jagger, W.S., Kaplan, M.W. and Bargoot, F.G. (1974), *Nature,* **251**, 31.
Lipton, S.A., Rasmussen, H. and Dowling, J.E. (1977), *J. gen. Physiol.*, **70**, 771.
Litman, B.J. (1974), *Biochemistry,* **13**, 2844.
Lowry, O.H., Roberts, N.R. and Lewis, G. (1956), *J. biol. Chem.*, **220**, 879.
Lowry, O.H., Roberts, N.R., Schultz, D.W., Clow, J.E. and Clark, J.R. (1961), *J. biol. Chem.*, **236**, 2813.
Marchesi, V.T., Furthmayr, H. and Tomita, M. (1976), *A. Rev. Biochem.*, **45**, 667.
Miki, N., Keirns, J.J., Marcus, F.R., Freeman, J. and Bitensky, M.W. (1973), *Proc. natn. Acad. Sci. U.S.A.*, **70**, 3820.
Miller, J.A., Brodie, A.E. and Bownds, M.D. (1975), *FEBS Letters,* **59**, 20.
Miller, J.A. and Paulsen, R. (1975), *J. biol. Chem.*, **250**, 4427.
Miller, J.E., Paulsen, R. and Bownds, M.D. (1977), *Biochemistry,* **16**, 2633.
Montal, M., Darszon, A. and Trissl, H.W. (1977), *Nature,* **267**, 221.
Montal, M. and Korenbrot, J.I. (1976), In: *The Enzymes of Biological Membranes* (Martonosi, A., ed.), Vol. 4, p. 365, Plenum Press, New York.
Nilsson, S.E.G. (1964), *J. Ultrastruct. Res.*, **11**, 581.
O'Brien, P.J. (1976), *Exp. Eye Res.*, **23**, 127.
O'Brien, P.J. (1977a), *Biochemistry,* **16**, 953.
O'Brien, P.J. (1977b), *Exp. Eye Res.*, **24**, 449.
O'Brien, P.J. (1978), *Exp. Eye Res.*, **26**, 197.
O'Brien, P.J. and Muellenberg, C.G. (1973), *Arch. Biochem. Biophys.*, **158**, 36.
O'Brien, P.J. and Muellenberg, C.G. (1974), *Exp. Eye Res.*, **18**, 241.
O'Brien, P.J. and Muellenberg, C.G. (1975), *Biochemistry,* **14**, 1695.
O'Brien, P.J., Muellenberg, C.G. and Bungenberg de Jong., J.J. (1972), *Biochemistry,* **11**, 64.
O'Brien, P.J. and Neufeld, E.F. (1972), In: *Glycoproteins: Their Composition, Structure and Function* (Gottschalk, A., ed.), 2nd edn., p. 1170, Elsevier, Amsterdam.
O'Day, W.T. and Young, R.W. (1978), *J. Cell. Biol.*, **76**, 593.

Ostroy, S.E. (1977), *Biochim. biophys. Acta,* **463**, 91.
Palade, G. (1975), *Science,* **189**, 347.
Papermaster, D.S., Converse, C.A. and Siu, J. (1975), *Biochemistry,* **14**, 1343.
Papermaster, D.S., Converse, C.A. and Zorn, M. (1976), *Exp. Eye Res.,* **23**, 105.
Papermaster, D.S. and Dreyer, W.J. (1974), *Biochemistry,* **13**, 2438.
Penn, R.D. and Hagins, W.A. (1972), *Biophys., J.,* **12**, 1073.
Plantner, J.J. and Kean, E.L. (1976), *J. biol. Chem.,* **251**, 1548.
Poo, M.-M. and Cone, R.A. (1973), *Exp. Eye Res.,* **17**, 503.
Poo, M.-M. and Cone, R.A. (1974), *Nature,* **247**, 438.
Raubach, R.A., Nemes, P.P. and Dratz, E.A. (1974), *Exp. Eye Res.,* **18**, 1.
Renthal, R., Steinemann, A. and Stryer, L. (1973), *Exp. Eye Res.,* **17**, 511.
Röhlich, P. (1976), *Nature,* **263**, 789.
Roth, S., McGuire, E.J. and Roseman, S. (1971a), *J. Cell Biol.,* **51**, 525.
Roth, S., McGuire, E.J. and Roseman, S. (1971b), *J. Cell Biol.,* **51**, 536.
Rothman, J.E. and Lenard, J. (1977), *Science,* **195**, 743.
Rothman, J.E. and Lodish, H.F. (1977), *Nature,* **269**, 775.
Saari, J.C. (1974), *J. Cell Biol.,* **63**, 480.
Saari, J.C. and Futterman, S. (1976), *Biochim. biophys. Acta,* **444**, 789.
Sack, R.A. and Harris, C.M. (1977), *Nature,* **265**, 465.
Saibil, H., Chabre, M. and Worcester, D. (1976), *Nature,* **262**, 266.
Schnetkamp, P.P.M., Daemen, F.J.M. and Bonting, S.L. (1977), *Biochim. biophys. Acta,* **468**, 259.
Schultze, M. (1866), *Arch. Mikr. Anat.,* **2**, 175.
Shichi, H., Somers, R.L. and O'Brien, P.J. (1974), *Biochem. biophys. Res. Commun.,* **61**, 217.
Shichi, H., Lewis, M.S., Irrevere, F. and Stone, A. (1969), *J. biol. Chem.,* **224**, 529.
Sitaramayya, A., Virmaux, N. and Mandel, P. (1977), *Exp. Eye Res.,* **25**, 163.
Sjostrand, F.S. (1953), *J. Cell. comp. Physiol.,* **42**, 15.
Smith, H.G., Fager, R.S. and Litman, B.J. (1977), *Biochemistry,* **16**, 1399.
Steinmann, A. and Stryer, L. (1973), *Biochemistry,* **12**, 1499.
Szuts, E.Z. and Cone, R.A. (1977), *Biochim. biophys. Acta,* **468**, 194.
Trayhurn, P., Habgood, J.O. and Virmaux, N. (1975), *Exp. Eye Res.,* **20**, 479.
Uhl, R., Hofmann, K.P. and Kreutz, W. (1977), *Biochim. biophys. Acta,* **469**, 113.
van Breugal, P.J.G.M., Daemen, F.J.M. and Bonting, S.L. (1975), *Exp. Eye Res.,* **21**, 315.
Waechter, C.J. and Lennarz, W.J. (1976), *A. Rev. Biochem.,* **45**, 95.
Wald, G. (1968), *Science,* **162**, 230.
Weidman, T.A. and Kuwabara, T. (1969), *Invest. Ophthal.,* **8**, 60.
Weller, M., Virmaux, N. and Mandel, P. (1975a), *Proc. natn. Acad. Sci. U.S.A.,* **72**, 381.
Weller, M., Virmaux, N. and Mandel, P. (1975b), *Nature,* **256**, 68.
Weller, M., Virmaux, N. and Mandel, P. (1976), *Exp. Eye Res.,* **23**, 65.
Wheeler, G.L. and Bitensky, M.W. (1977), *Proc. natn. Acad. Sci. U.S.A.,* **74**, 4238.

Wiggert, B.O. and Chader, G.J. (1975), *Exp. Eye Res.*, **21**, 143.
Wirtz, K.W.A. (1974), *Biochim. Biophys. Acta,* **344**, 95.
Worthington, C.R. (1974), *A. Rev. Biophys. Bioeng.*, **3**, 53.
Wu, C-W. and Stryer, L. (1972), *Proc. natn. Acad. Sci. U.S.A.*, **69**, 1104.
Young, R.W. (1967), *J. Cell Biol.*, **33**, 61.
Young, R.W. (1968), *J. Ultrastruct. Res.*, **23**, 462.
Young, R.W. (1969), *Invest. Ophthal.*, **8**, 222.
Young, R.W. (1971), *J. Cell Biol.*, **49**, 303.
Young, R.W. (1975), *Vision Res.*, **15**, 535.
Young, R.W. (1976), *Invest Ophthal.*, **15**, 700.
Young, R.W. (1977), *J. Ultrastruct. Res.*, **61**, 172.
Young, R.W. (1978), *Invest. Ophthal.*, **17**, 105.
Young, R.W. and Bok, D. (1969), *J. Cell Biol.*, **42**, 392.
Young, R.W. and Droz, B. (1968), *J. Cell Biol.*, **39**, 169.

Author Index to Series A, Volumes One to Six

Numbers in parentheses are volume references

Subject Index to Series A, Volumes One to Six

Numbers in parentheses are volume references.